Alan Smith

Revision Guide

For the Cambridge Secondary 1 Test

Answers can be found at www.hoddereducation.com/checkpointextras

This text has not been through the Cambridge endorsement process.

Orders: please contact Bookpoint Ltd, 130 Milton Park, Abingdon, Oxon OX14 4SB. Telephone: (44) 01235 827827. Fax: (44) 01235 400401. Lines are open 9.00–5.00, Monday to Saturday, with a 24-hour message answering service. Visit our website at www.hoddereducation.com.

First published in 2013 by
Hodder Education
An Hachette UK Company
London NW1 3BH

Impression number	5	4	3	2	1
Year	2015		2014		2013

Illustrations by Gray Publishing
Typeset in 12/14pt Garamond and produced by Gray Publishing, Tunbridge Wells
Printed in Spain

A catalogue record for this title is available from the British Library

ISBN 978 1444 18071 8

Contents

The chapters in this book have been arranged to match the Cambridge Secondary 1 Mathematics Curriculum Framework as follows:

- Number
- Algebra
- Geometry
- Measure
- Handling data
- Calculation and mental strategies
- Problem solving

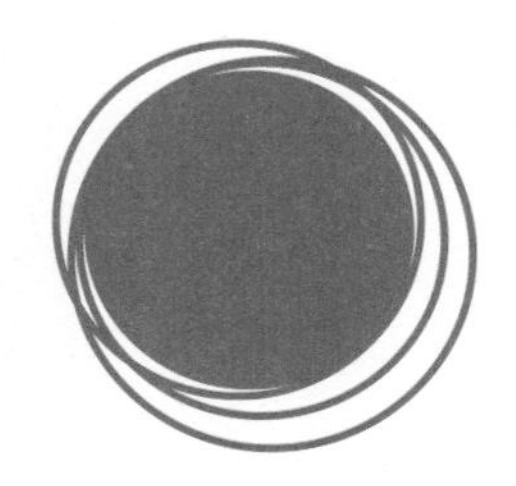

Introduction

Preparing for the Test

You may find the following points helpful in preparing for the Cambridge Checkpoint Mathematics, Cambridge Secondary 1 Test.

- Make sure you are familiar with the mathematical content that the test will cover. Use this revision guide to check your understanding of each of the topic areas.
- Read the Tips for success for some useful pointers of what the test is looking to check your understanding on.
- Make sure you are familiar with the types of questions you will be asked in the test.
- Obtain some past papers, and practise working through these in the time limits given, so that you know what is expected of you.
- Spend some time practising the test-style questions (Spotlight on the test) in this revision guide.
- Remember to look at the mark scheme so that you know how much each answer is worth.
- Check your answers against the sample answers supplied so that you can see how your answers will be assessed. You can download the answers for free at www.hoddereducation.com/checkpointextras

General revision tips

- Find somewhere quiet to revise. Sit on a comfortable chair at a table and have a pen or pencil and some sheets of paper as well as this book. Plan what you will revise in your revision session. Remember to take a break perhaps every twenty minutes or half an hour to let your mind rest.
- When using a calculator, write down the calculation you are intending to do first, before entering it onto the calculator keypad.
- Just reading through the text is not always the best way of learning. It is better to make your revision more active. You should use a variety of active ways to make your learning secure. Here are some of them:
  - Cover up the solution to a worked example on the topic you are revising. Write out your own solution, and then check it against the worked solution. In addition to getting the right answer, make sure all your key steps of working are shown clearly.
  - Continue your active revision by completing the Spotlight on the test sections. Write your answers on paper.
  - Study the Tips for success; write up some of them on revision cards.

Place value, ordering and rounding

Comparing two quantities

Student's book references

- Book 1 Chapter 1
- Book 2 Chapter 1
- Book 1 Chapter 8

Worked example

Insert a > or < sign to show the larger number.

6324 ____ 6234

Solution

Comparing the highest place values in the two numbers:

6324
6234

Both numbers contain 6000, so now compare the next place value:

6**3**24
6**2**34

The first number contains 300, the second only 200.
Thus 6324 is greater than 6234, and we write this as:

6324 > 6234

Tips for success

- Remember that the sign > means 'is greater than'.
- The sign < means 'is less than'.
- If you forget which is which – the larger quantity goes at the wider end of the symbol.

Check your understanding 1.1

Insert a > or < sign between each pair of numbers.

1 623 ____ 652
2 3108 ____ 3112
3 0.235 ____ 0.215
4 9740 ____ 12 350
5 13.226 ____ 12.895

Multiplying and dividing by powers of 10

If you multiply or divide a number by 10, all of the place values shift one position. Multiplying or dividing by 100 moves them two positions, and so on.

Worked examples

Work out:

a) 2992×100

b) $38\,400 \div 10$

Solution

a) $2992 \times 100 = 292\,200$ (two position shift).

b) $38\,400 \div 10 = 3840$ (one position shift).

Check your understanding 1.2

Work out each of these multiplications or divisions.

1 362×10
2 $13700 \div 10$
3 1220×100
4 $14000 \div 100$
5 $1800 \div 10$
6 600×100
7 740×1000
8 $300000 \div 100$
9 $540000 \div 1000$
10 130×100

Rounding whole numbers

Sometimes numbers are recorded to a higher level of accuracy than we really need. In the case of whole numbers we often then round them to the nearest ten, hundred or thousand.

Worked example

A school has 785 pupils in the Middle School years, and 215 in the Sixth Form. Round these numbers, to the nearest hundred in each case.

Solution

Rounding to the nearest hundred means we can only use values like 600, 700, 800, 900:

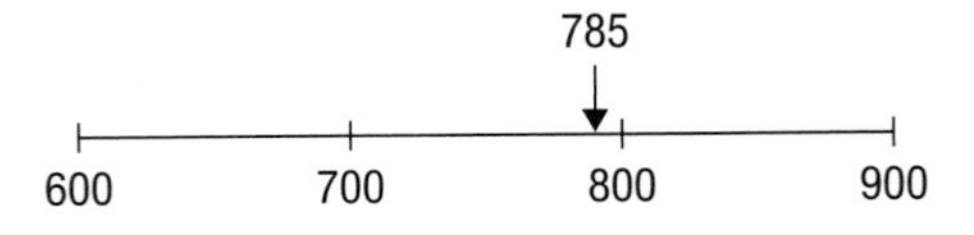

Clearly 785 is nearer to 800 than any of these other options. So, 785 to the nearest hundred is 800 (rounded up). Similarly, 215 to the nearest hundred is 200 (rounded down).

Tips for success

- If you are exactly halfway between two options, round up, not down. So, for example, 450 would round to 500, not 400.

Check your understanding 1.3

Round each of these as indicated.

1 2532 (nearest 10).
2 4760 (nearest 100).
3 87 (nearest 10).
4 259 (nearest 10).
5 259 (nearest 100).
6 6822 (nearest 10).
7 A football match is watched by 12 371 people. Round this number to the nearest hundred.
8 A house costs $123 499 to build. Round this cost to the nearest thousand dollars.

Decimal places

Some numbers may have a large number of decimal places, and we may wish to round them to fewer places. This follows a similar procedure to that for whole numbers in the previous section.

Tips for success

- If the first figure after the cut lies between 0 and 4 you round down.
- If this figure lies between 5 and 9 you round up.
- In the worked example part b) it would have been wrong to write 0.17 – even though this has the same value as 0.170 – because then you are rounding to two decimal places.

Worked examples

Round these numbers to three decimal places:

a) 16.2371 **b)** 0.169 73.

Solution

a) 16.237**1**. The first figure after three decimal places is low – 1 – so we round down to 16.237.

b) 0.169**7**3. The first figure after three decimal places is high – 7 – so we round up to 0.170.

Check your understanding 1.4

Round each of these as indicated.

1 18.562 (one decimal place).
2 304.849 (one decimal place).
3 8.0671 (two decimal places).
4 28.2219 (three decimal places).
5 61.4587 (two decimal places).
6 72.203 (two decimal places).
7 3.141 59 (three decimal places).
8 1.414 (one decimal place).
9 0.0688 (two decimal places).
10 7.0707 (three decimal places).

Significant figures

Significant figures are the figures that give information about the size of each of the place values. Significant figures are counted from the left. For example, 3285 contains four significant figures (s.f.). The number 0.002 16 has three s.f., since the 2 is the first significant figure counting from the left.

Tips for success

- Sometimes zeros count as significant figures; other times they do not (they are just place holders). Study these examples carefully to make sure you understand the different ways in which zeros are used.
- In the worked example part d) it would have been wrong to write 0.069 – even though this has the same value as 0.0690 – because then you are rounding to two significant figures.

Worked examples

Write these numbers to three significant figures:

a) 19 345 **b)** 306 820 **c)** 0.006 086 **d)** 0.068 955.

Solution

a) 19 3**4**5 rounds down, to become 19 300.

b) 306 **8**20 rounds up, to become 307 000.

c) 0.006 08**6** rounds up, to become 0.00 609.

d) 0.068 9**5**5 rounds up, to become 0.0690.

Check your understanding 1.5

Round each of these as indicated.

1 3.141 59 (two significant figures).
2 156.13 (four significant figures).
3 165.5 (three significant figures).
4 154 285 (four significant figures).
5 16 324 (three significant figures).
6 851 (one significant figure).
7 2524 (three significant figure).
8 0.003 16 (two significant figures).
9 0.010 13 (two significant figures).
10 1.357 (one significant figure).

Estimating the answer to a calculation

To obtain an estimate of the answer to a multiplication or division problem, you can round off all of the quantities to one significant figure, and then do the calculation using these approximate values.

Worked example

There are 32 coaches on a cross-river ferry. Each coach carries 48 passengers. Estimate the total number of passengers on the coaches.

Solution
The exact calculation is 32×48.
The approximate calculation is $30 \times 50 = 1500$.
So the coaches carry approximately 1500 people in total.

Tips for success

- Do not try to be too precise here: the purpose of the estimate is to get an answer to one significant figure only. There is no need to do an exact calculation.

Check your understanding 1.6

Round the numbers to one significant figure and hence make estimates of the answers to each of these.

1 Estimate the answer to 231×39.
2 Estimate the answer to 52×22.
3 Estimate the answer to $1721 \div 48$.
4 Estimate the answer to $611 \div 28$.
5 A chocolate bar weighs 65 grams. Find the total weight of 18 bars.
6 463 sweets are shared out between 21 people. Find out how many sweets they each get.
7 Some wedding guests travel to a reception by taxi. Each taxi can carry four guests. How many taxis are needed for 59 people?
8 39 215 toy bricks are sorted into bags, each containing 200 bricks. How many bags will there be?
9 A group of 32 friends win \$613 000. How much will they each receive, if they share it out equally among themselves?
10 Baby William has 19 boxes of pencils; each box contains 12 pencils. He shakes them all out on the floor. How many pencils are there in total?

Spotlight on the test

1 Insert one of the symbols >, < or = into the middle of each statement.

a) 7325 _____ 7236.

b) 20×1000 _____ 200×100.

c) 29×59 _____ 31×61.

d) $40\,000 \div 100$ _____ $400 \div 10$. [4]

2 Insert the missing number into this calculation:

$0.6221 \times$ _____ $= 6221$. [1]

3 Using the digits 2, 5, 8, 9 once each, make a four-digit number that meets all of these rules:

- The number is greater than 9000.
- The number is odd.
- If rounded to the nearest thousand, the number rounds down. [1]

4 b) Work out the answer to $\frac{8.3 \times 4.7}{2.5}$ using a calculator. Write down all the figures.

b) Now write your answer to part a) correct to three significant figures. [2]

5 Look at the numbers on these cards:

285.3	21.03

a) Write each number correct to one significant figure.

b) Without working out the exact answer, make an estimate for the value of 285.3×21.03 correct to one significant figure. Write down all of your working. [3]

6 The number of seconds in one year can be worked out like this:

$365 \times 24 \times 60 \times 60$.

a) Explain where each of these numbers comes from.

b) Estimate the answer to 365×24 by rounding the numbers to one significant figure.

c) Work out the exact answer to 60×60. Give your answer to one significant figure.

d) Using your answers to b) and c) estimate the number of seconds in one year. Give your final answer correct to one significant figure. [4]

2 Integers, powers and roots

Adding, subtracting and ordering with positive and negative numbers

Student's book references

- Book 1 Chapter 8
- Book 2 Chapter 8
- Book 3 Chapters 1 and 8

The integers are the positive and negative whole numbers, and zero. Addition and subtraction problems with integers are often illustrated using the number line. Adding a positive number corresponds to moving to the right, and subtracting a positive number means moving to the left. Adding or subtracting a negative integer causes these rules to reverse.

Worked examples

Use the number line to work out:

a) (+5) + (–2) **b)** (–3) – (+6) **c)** (+6) – (–2).

Solution

a) For (+5) + (–2), start at (+5) and move two in the negative direction:

–10 –8 –6 –4 –2 0 2 4 6 8 10

So (+5) + (–2) = (+3) = 3.

b) For (–3) – (+6) start at (–3) and move six in the negative direction:

–10 –8 –6 –4 –2 0 2 4 6 8 10

So (–3) –(+ 6) = (–9) = –9.

c) For (+6) – (–2) start at (+6) and move two in the positive direction:

–10 –8 –6 –4 –2 0 2 4 6 8 10

So (+6) – (–2) = (+8) = 8.

The same principles can be applied to addition and subtraction problems with decimal numbers.

Worked example

Without a calculator, find the value of (2.4) + (–3.7).

Solution

(2.4) + (–3.7) is the same as 2.4 – 3.7 = –1.3.

Check your understanding 2.1

Sketch a number line from –10 to +10 and use it to help answer these questions.

1 a) (+3) + (+2) b) (+8) + (–1) c) (+6) + (–5).
2 a) (–3) + (–4) b) (–6) + (+2) c) (+5) + (–7).
3 a) (+6) – (+2) b) (–3) – (–1) c) (–6) – (+1).
4 a) (–5) + (–5) b) (–1) – (+1) c) (+3) + (–4).
5 a) (+6.8) – (+7.2) b) (–3.1) + (+1.5) c) (–4.2) – (–4.2).
6 a) (+4.5) + (–3.8) b) (–2.2) + (–1.8) c) (–2.1) – (–6.2).
7 a) (+7.2) – (–7.2) b) (+7.2) + (–7.2) c) (–1.4) – (+2.5).
8 Arrange these in order of size, smallest first: (–4), (+6), (–3).
9 Arrange these in order of size, smallest first: (–1), (+2), (–5), (–10).
10 Arrange these in order of size, smallest first: (+1), (–8), (–5), (–10).

Factors, multiples, primes and tests of divisibility

Tips for success

- Remember that if a number is divisible by (for example) 5, then it is called a **multiple** of 5. We say 5 is a **factor** of the number.
- A number which has no factors (other than 1 and itself) is called a **prime** number. The first few primes are 2, 3, 5, 7, 11 …

Property	Test	Example
A number is divisible by 2	The units digit is 0, 2, 4, 6 or 8	5**6**
A number is divisible by 3	The sum of the digits is divisible by 3	87 as 8 + 7 = 15 = 5 × **3**
A number is divisible by 5	The units digit is 0 or 5	4**5**
A number is divisible by 6	The number is divisible by 2 and 3	42 as 4**2** and 4 + **2** = 6 = 2 × **3**
A number is divisible by 9	The sum of the digits is divisible by 9	774 as 7 + 7 + 4 = 18 = 2 × **9**

Check your understanding 2.2

1 Look at this list of whole numbers (integers): 10, 11, 12, 13, 14, 15, 16, 17
 a) Which of the numbers are divisible by 3?
 b) Which are divisible by 5?
 c) Which numbers are prime?
2 Find the largest number which is a factor of 35 and also a factor of 65. (This is called the highest common factor of the two numbers.)
3 Find the highest common factor of 18 and 30.
4 Here is part of a sieve of Eratosthenes. Prime numbers are marked in bold and shaded. Non-primes are greyed out.

 Use the sieve to answer the following questions.

1	**2**	**3**	4	**5**	6	**7**	8	9	10
11	12	**13**	14	15	16	**17**	18	**19**	20
21	22	**23**	24	25	26	27	28	**29**	30
31	32	33	34	35	36	**37**	38	39	40
41	42	**43**	44	45	46	**47**	48	49	50
51	52	**53**	54	55	56	57	58	**59**	60
61	62	63	64	65	66	**67**	68	69	70
71	72	**73**	74	75	76	77	78	**79**	80

 a) Find the next prime above 31.
 b) Write down all the primes between 40 and 50.
 c) Is it true that 51, 61 and 71 are all primes?
 d) How many primes are there between 1 and 20?

Multiplication and division of integers

To multiply or divide signed integers, carry out the multiplication or division without any signs first. Then count the number of minus signs. No minus signs: positive; one minus sign: negative; two minus signs: positive (and so on if needed).

Worked examples

Work out the values of **a)** $(+5) \times (-2)$, **b)** $(-6) \div (-2)$.

a) For $(+5) \times (-2)$, we know that $5 \times 2 = 10$. One minus sign means a negative answer, so $(+5) \times (-2) = (-10) = -10$.

b) For $(-6) \div (-2)$, we know that $6 \div 2 = 3$. Two minus signs means a positive answer, so $(-6) \div (-2) = (+3) = 3$.

Check your understanding 2.3

1 Work out the answers to these multiplications:
a) $(+7) \times (-2)$ b) $(-3) \times (+4)$ c) $(+2) \times (+9)$ d) $(-2) \times (-9)$.

2 Work out the answers to these divisions:
a) $(-10) \div (+2)$ b) $(+20) \div (-5)$ c) $(-16) \div (-8)$ d) $(+30) \div (-2)$.

3 Work out the answers to these multiplications and divisions:
a) $(+8) \times (-6)$ b) $(+20) \div (-5)$ c) $(-12) \times (-3)$ d) $(-40) \div (-4)$.

Factors and factor trees

Factors of a number are all the whole numbers (positive integers) that divide exactly into that number.

- The factors of 20 are 1, 2, 4, 5, 10 and 20.
- The factors of 30 are 1, 2, 3, 5, 6, 10, 15 and 30.

The largest factor to appear in both groups is 10, so 10 is the highest common factor (HCF) of 20 and 30.

- The multiples of 6 are those numbers in the 6× table: 6, 12, 18, 24, 30, and so on.
- The multiples of 8 are those numbers in the 8× table: 8, 16, 24, 32, 40, and so on.

The smallest multiple to appear in both groups is 24, so 24 is the lowest common multiple (LCM) of 6 and 8.

Worked examples

a) Write 48 and 60 as products of their prime factors.

b) i) Find the highest common factor (HCF) and

ii) the lowest common multiple (LCM) of 48 and 60.

Solution

a) 48 and 60 may be factorised using factor trees (or the table method):

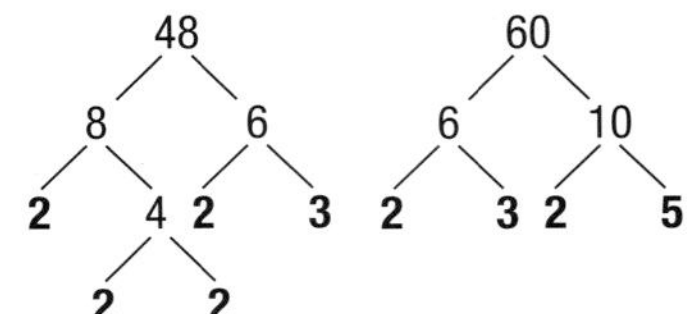

So $48 = 2^4 \times 3$ and $60 = 2^2 \times 3 \times 5$.

b) The factors may be represented in a Venn diagram:

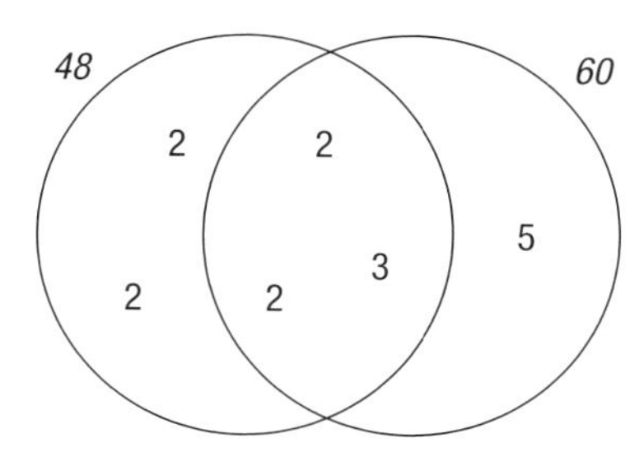

i) To find the HCF, look at the overlap (intersection) of the two circles: HCF (48, 60) = $2 \times 2 \times 3 = 12$.

ii) To find the LCM, look at the entire diagram: LCM (48, 60) = $2^4 \times 3 \times 5 = 240$.

Check your understanding 2.4

1 Write each of these as a product of primes. You may use a factor tree to help.
a) 45 b) 44 c) 72 d) 180.

2 Work out the highest common factor (HCF) of 28 and 42.

3 Work out the lowest common multiple (LCM) of 35 and 45.

4 a) Express 120 and 144 in prime factor form.
b) Find the highest common factor (HCF) of 120 and 144.

5 a) Express 75 and 120 as in prime factor form.
b) Find the lowest common multiple (LCM) of 75 and 120.

Squares, cubes and roots

Here are some of the more commonly encountered squares and cubes:

The first 20 square numbers									
1^2	2^2	3^2	4^2	5^2	6^2	7^2	8^2	9^2	10^2
1	4	9	16	25	36	49	64	81	100
11^2	12^2	13^2	14^2	15^2	16^2	17^2	18^2	19^2	20^2
121	144	169	196	225	256	289	324	361	400

The first ten cube numbers									
1^3	2^3	3^3	4^3	5^3	6^3	7^3	8^3	9^3	10^3
1	8	27	64	125	216	343	512	729	1000

Worked examples

Write down the values of **a)** 14^2, **b)** 6^3, **c)** $\sqrt{196}$, **d)** $\sqrt[3]{343}$

Solution

a) $14^2 = 196$

b) $6^3 = 216$

c) $\sqrt{324} = 18$ (because $= 18^2 = 324$)

d) $\sqrt[3]{343} = 7$ (because $7^3 = 343$).

Worked example

The square root of 73 lies between the integers n and $n + 1$. Find the value of n.

Solution

The square numbers 64 and 81 are located below and above 73, so

$$64 < 73 < 81$$
$$\sqrt{64} < \sqrt{73} < \sqrt{81}$$
$$8 < \sqrt{73} < 9$$

Therefore $n = 8$.

Check your understanding 2.5

1 Write down the values of **a)** 11^2, **b)** 6^3, **c)** $\sqrt{144}$, **d)** $\sqrt[3]{64}$
2 Write down the values of **a)** 9^2, **b)** 8^3, **c)** $\sqrt{289}$, **d)** $\sqrt[3]{729}$
3 Write down the values of **a)** 13^2, **b)** 5^3, **c)** $\sqrt{361}$, **d)** $\sqrt[3]{27}$
4 Here are some clues about a number:
- The number is a perfect cube.
- The number has three digits.
- The digits multiply together to make 36.

What is the number?
5 Explain how you can tell whether the square root of 105 lies between 10 and 11.
6 Pedro says that the square root of 43 lies between 5 and 6.
Anton says that the square root of 43 lies between 6 and 7.
Explain how you can tell which one of them is right.

Working with indices

Quantities involving repeated multiplication can be written using index notation, such as

$$2 \times 2 \times 2 \times 2 \times 2 = 2^5$$

In this example the number 2 is the **base** and 5 is the **index** (or power). Here are some commonly used laws of indices:

Law	Example
$a^m \times a^n = a^{m+n}$	$5^3 \times 5^4 = 5^7$
$a^m \div a^n = a^{m-n}$	$2^8 \div 2^5 = 2^3$
$(a^m)^n = a^{mn}$	$(4^3)^5 = 4^{15}$
$x^{-n} = \frac{1}{x^n}$	$5^{-2} = \frac{1}{5^2} = \frac{1}{25}$
$x^0 = 1$	$7^0 = 1$

Worked example

Write as a single power $4^4 \times 4^5 \div 4^3$

Solution

$$\begin{aligned} 4^4 \times 4^5 \div 4^3 &= 4^{4+5} \div 4^3 \\ &= 4^9 \div 4^3 \\ &= 4^{9-3} \\ &= 4^6. \end{aligned}$$

Check your understanding 2.6

Write each of these as a single power:

1 $3^4 \times 3^5$

2 $6^2 \times 6^3$

3 $4^5 \div 4^2$

4 $2^7 \div 2^5$

5 $5^4 \times 5^8$

6 $4^4 \div 4^5$

7 $(8^4)^3$

8 $(5^3)^2$

9 $4^4 \times 4$

10 $10^6 \div 10^5$

11 $6^7 \times 6^{-2}$

12 $7^8 \div 7^5$

13 Simplify $3^5 \times 3^4$

14 Find the value of 4^0

15 Find the value of m if $2^4 \times 2^m = 2^7$

16 Find the value of t if $4^5 \div 4^t = 4^8$

Order of operations

Tips for success

- Divide and multiply have equal priority to each other. If both are present, do them in the order you meet them.
- Addition and subtraction also have equal priority (but lower than multiplication and division).
- Modern calculators have BIDMAS programmed in automatically.

Some calculations involve several different operations. Brackets should always be worked out first, followed by indices. Then come divisions and multiplications, and finally additions and subtractions. These are conveniently remembered as BIDMAS:

B brackets

I indices

D division

M multiplication

A addition

S subtraction

Worked example

Find the value of $5 \times 6 + 7 \times 8$.

Solution

The multiplications are done before adding, so

$$\begin{aligned} 5 \times 6 + 7 \times 8 &= 30 + 56 \\ &= 86. \end{aligned}$$

Worked example

Find the value of $15 + (2 + 4)^2$.

Solution

$$
\begin{aligned}
15 + (2 + 4)^2 &= 15 + (6)^2 && \text{(brackets)}\\
&= 15 + 36 && \text{(indices)}\\
&= 51 && \text{(addition).}
\end{aligned}
$$

Check your understanding 2.7

Find the value of each of these. Do not use your calculator.

1 $3^2 + 5^2$
2 $(3 + 5)^2$
3 $5 + 3 \times 2$
4 $14 \div (5 + 2)$
5 $(3 + 4) \times (5 + 6)$
6 $3 + 4 \times 5 + 6$
7 $3 \times (30 - 5^2)$
8 $(18 + 42) \div (27 - 7)$
9 $(5^2 - 3^2) \div 2$
10 $8 \times 6 \div 12$

Spotlight on the test

1 Look at this list of integers:

20, 21, 22, 23, 24, 25, 26, 27, 28

a) Which one is a multiple of 6?
b) Which one is a factor of 60?
c) Which one is prime? [3]

2 On the Moon the daytime temperature reaches 105 °C, but at night falls to −155 °C. Work out the difference between the daytime and night temperatures on the Moon. [2]

3 Adam the gardener mows his lawns every six days. He weeds the flower beds once every eight days. On 1 July he mows his lawns and weeds the flower beds. On what date does he next do both tasks on the same day? [2]

4 During one week last year I recorded the temperatures as 2 °C on Tuesday, −6 °C on Wednesday and −4 °C on Thursday.
a) Which day was coldest?
b) How much warmer was Tuesday compared with Thursday? [2]

5 Fred says: 'The square root of 60 is between 7 and 8.'
Evie says: 'The cube root of 250 lies between 7 and 8.'
a) Explain how you can tell that Fred is right.
b) Explain how you can tell that Evie is wrong. [2]

6 Work out the values of a, b and c in these statements:
a) $5^3 \times 5^a = 5^7$.
b) $7^5 \div 7^8 = 7^b$.
c) $9^c = 1$. [3]

7 Work out the value of $6 + 7 \times 5$. [1]

8 Simplify: **a)** $(5^3)^4$, **b)** $2^3 \times 2^5$, **c)** 6^0. [3]

9 Find the value of $10 + 4 \times 3^2$. [2]

10 The lowest common multiple of 10 and y is 60. Find the value of y (given that y is not 60). [1]

Expressions, equations and formulae

Expressions and equations

Student's book references

- Book 1 Chapter 2
- Book 2 Chapter 2
- Book 3 Chapter 2

- An **expression** is a piece of algebra such as $2x + 1$. This example contains one **variable**, x.
- An expression may have several **terms**, separated by + or – signs. $2x + 1$ has two terms.
- An **equation** contains an equals (=) sign, such as $x + 1 = 10$. This equation is true when $x = 9$ and is not true for all other values of x; we say that $x = 9$ is a **solution** of the equation.
- A **formula** is like an equation, but often contains more than one variable. You cannot solve a formula; rather, it is used to work out the value of one quantity if you know all the others. $V = \pi r^2 h$ is an example of a formula.

Check your understanding 3.1

Say whether each of these is best described as an expression, an equation or a formula.

1 $5x + 3$
2 $V = I \times R$
3 $4x + 1 = 33$
4 $x^2 + 7x - 3$
5 $E = mc^2$
6 πr^2
7 $C = 2\pi r$
8 $y + 3x$
9 $x^2 = 144$
10 $c^2 = a^2 + b^2$

Working with indices

Here is a reminder of the laws of indices, with some examples:

Law	Example
$x^a \times x^b = x^{a+b}$	$x^3 \times x^4 = x^7$
$x^a \div x^b = x^{a-b}$	$x^8 \div x^5 = x^3$
$(x^a)^n = x^{an}$	$(x^3)^5 = x^{15}$
$x^{-n} = \frac{1}{x^n}$	$x^{-2} = \frac{1}{x^2}$
$x^0 = 1$	

Check your understanding 3.2

Simplify each of these expressions, which have been written using indices.

1 $x^6 \times x^3$
2 $y^6 \times y^{-2}$
3 $z^{10} \div z^4$
4 $x^2 \times x^5$
5 $y^8 \div y^4$
6 $z^7 \div z^7$
7 $y^7 \div y^4$
8 $(z^3)^3$
9 $m^3 \div m^5$
10 $z^4 \times z^{-2}$

Collecting like terms, and expanding brackets

If an expression contains several terms of the same kind, these can be collected together ('simplified') into a single term.

Worked example

Simplify $10x + 7y - 3x + 2y$.

Solution

$$10x + 7y - 3x + 2y = 10x - 3x + 7y + 2y$$
$$= 7x + 9y$$

Some expressions contain **brackets**. It can be helpful to multiply out the brackets ('expand') and then collect up like terms in the result ('simplify').

Worked example

Simplify $3(a + 5b) + 2(a - 2b)$.

Solution

$$3(a + 5b) + 2(a - 2b) = 3a + 15b + 2a - 4b$$
$$= 5a + 11b$$

Special care needs to be taken when one bracket is subtracted from another.

Worked example

Simplify $4(2x + 3y) - 2(x - 2y)$.

Solution

$$4(2x + 3y) - 2(x - 2y) = 8x + 12y - 2x + 4y$$ (as -2 times $-2y$ is $+4y$)
$$= 6x + 16y$$

Check your understanding 3.3

Multiply out the brackets in these expressions.

1 $2(x + 4)$

2 $3(x + 2)$

3 $5(2x - 1)$

4 $7(a + 5)$

5 $10(2b - 3)$

6 $4(3x + 2y)$

7 $2(x - 3y)$

8 $4(2x + 3y)$

9 $9(2a - 3b)$

10 $10(2b + 5)$

Expand and simplify these bracketed expressions.

11 $4(x + 1) + 5(x + 2)$
12 $3(x + 2) + 2(x - 1)$
13 $4(5x + 2) + 3(8x + 3)$
14 $3(5y + 1) + 2(y - 4)$
15 $2(3x + 6y) - 3(x + 3y)$
16 $4(x - 1) - 2(x - 5)$
17 $6(m - 2) + 3(m + 4)$
18 $3(2n - 1) - 2(n + 1)$
19 $5(2k - 1) + 2(3k + 5)$
20 $7(p - 1) - 2(3p - 5)$

Factorising

Factorising is the reverse process of expanding. The idea is to look for a common factor and take it outside a set of brackets.

Worked example

Factorise $10x + 16$.

Solution
Notice that 2 is a factor of $10x$ and also of 16, so $10x + 16 = 2(5x + 8)$.

The factor to be removed may be a number, a letter or both.

Worked example

Factorise $5a^2 - 15a$.

Solution
Notice that $5a$ is a factor of $5a^2$ and also of $15a$, so $5a^2 - 15a = 5a(a - 3)$.

Check your understanding 3.4

Factorise the following.

1 $7x + 21$
2 $4y + 10$
3 $5z - 20$
4 $9x + 15$
5 $15y - 25$
6 $x^2 + 7x$
7 $3y^2 - 9y$
8 $10y^2 + 25y$
9 $12z^2 - 15z$
10 $4xy - 10x$

Changing the subject of a formula

A formula is a statement like $V = \pi r^2 h$. We say that V is the **subject** of the formula. This formula can be rearranged to state that $h = V/(\pi r^2)$ so that h is now the subject. This is called changing the subject of a formula.

Worked example

Make x the subject of the formula $y = 5x + 3$.

Solution

$$5x + 3 = y$$
$$5x = y - 3$$
$$x = \frac{y - 3}{5}$$

Some examples involve squares and square roots. Care must be taken to carry out the steps in the right order.

Worked example

Make c the subject of $E = mc^2$.

Solution

$$mc^2 = E$$
$$c^2 = \frac{E}{m}$$
$$c = \sqrt{\frac{E}{m}}$$

Tips for success

- We need to make c^2 the subject first, then take the square root to get c.

Check your understanding 3.5

Rearrange each formula, to make the given letter (in brackets) the subject.

1	$V = abc.$	(b)	**2**	$y = mx + c.$	(m)
3	$C = 2\pi r.$	(r)	**4**	$PV = RT.$	(T)
5	$W = 8h^2.$	(h)	**6**	$u = 2a + n.$	(n)
7	$u = 2a + n.$	(a)	**8**	$F = ma.$	(a)
9	$A = \pi r^2.$	(r)	**10**	$v^2 = u^2 + 2as.$	(u)

Adding and subtracting algebraic fractions

Algebraic fractions can be added or subtracted, provided the fractions have the same denominator.

Worked example

Simplify into a single fraction $\frac{x + 1}{4} - \frac{x}{6}$

Solution

$$\frac{x + 1}{4} - \frac{x}{6} = \frac{3(x + 1)}{12} - \frac{2x}{12}$$
$$= \frac{3(x + 1) - 2x}{12}$$
$$= \frac{3x + 3 - 2x}{12} = \frac{x + 3}{12}$$

Tips for success

- We need to write both fractions with the same denominator (12 in this case).

Check your understanding 3.6

Write each of these as a single fraction.

1 $\frac{1}{8}+\frac{1}{2}$

2 $\frac{x}{8}+\frac{x}{2}$

3 $\frac{3}{5}-\frac{1}{4}$

4 $\frac{3x}{5}-\frac{x}{4}$

5 $\frac{2y}{3}-\frac{3y}{7}$

6 $\frac{7a}{10}-\frac{a}{2}$

7 $\frac{2b}{3}-\frac{b}{10}$

8 $\frac{19c}{20}-\frac{3c}{5}$

9 $\frac{x}{8}+\frac{x+3}{2}$

10 $\frac{y+2}{8}+\frac{y+5}{2}$

11 $\frac{3x}{5}-\frac{1}{4}$

12 $\frac{7}{x}-\frac{3}{2x}$

Spotlight on the test

1 Decide whether each of these is best described as an expression, an equation or a formula:

a) $x^2 + 3x + 1$

b) $A = \pi r^2$

c) $2 + 5x = 3x$. [3]

2 Simplify $4x(x + 5)$. [2]

3 Expand and simplify: $(2x + 1)(x - 3)$. [2]

4 Factorise $6m^2 + 8mn$. [2]

5 Simplify **a)** $y^3 \times y^4$, **b)** $x^{10} \div x^5$, **c)** $(z^4)^2$. [3]

6 Make s the subject of $v^2 = u^2 + 2as$. [2]

7 Make x the subject of $y = 3x + 9$. [2]

8 Look at the expressions on the cards below:

$m^3 \div m^2$	m^0	$m^3 \times 1/m^2$	$m^{-3} \div m^{-4}$	$m^6/(m^3)^2$
A	**B**	**C**	**D**	**E**

a) Which of them is/are equal to 1?

b) Which of them is/are equal to m? [2]

9 Write as a single fraction: $\frac{5x}{8}+\frac{x}{4}$. [2]

10 Write as a single fraction: $\frac{x}{5}+\frac{x+1}{10}$. [2]

Shapes, congruency and geometric reasoning

Congruent shapes and triangles

Student's book references

- Book 1 Chapter 3
- Book 2 Chapter 3
- Book 3 Chapter 3

If two objects are the same shape and size then we say they are **congruent**. (If they are the same shape but one is larger than the other then they are **similar**.)

Two triangles will be congruent if they have three matching sides (SSS), two matching sides and an included angle (SAS), two matching angles and an included side (ASA) or matching hypotenuse and side in a right-angled triangle (RHS).

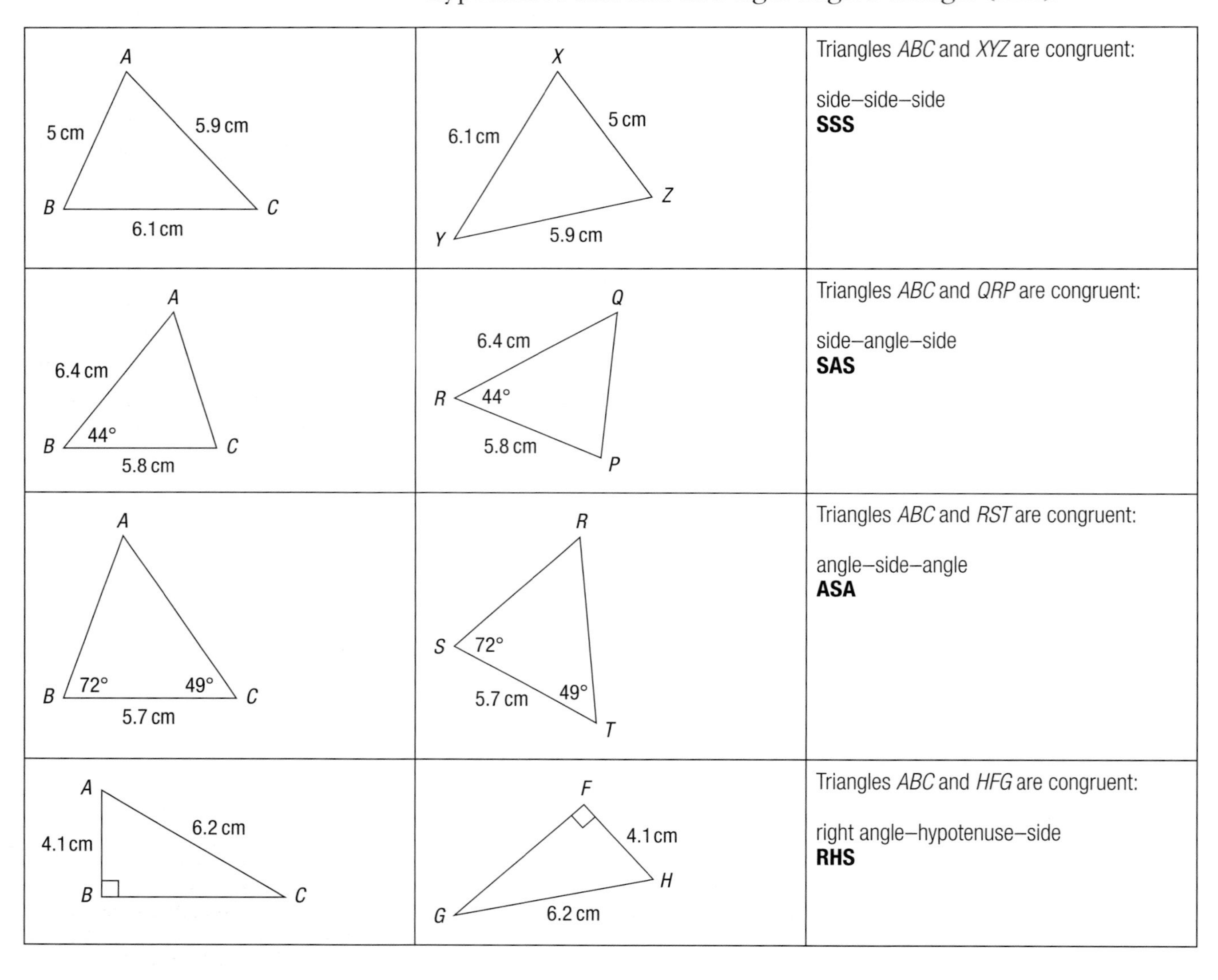

Check your understanding 4.1

State, with reasons, whether each pair of triangles in **1** to **4** is congruent.

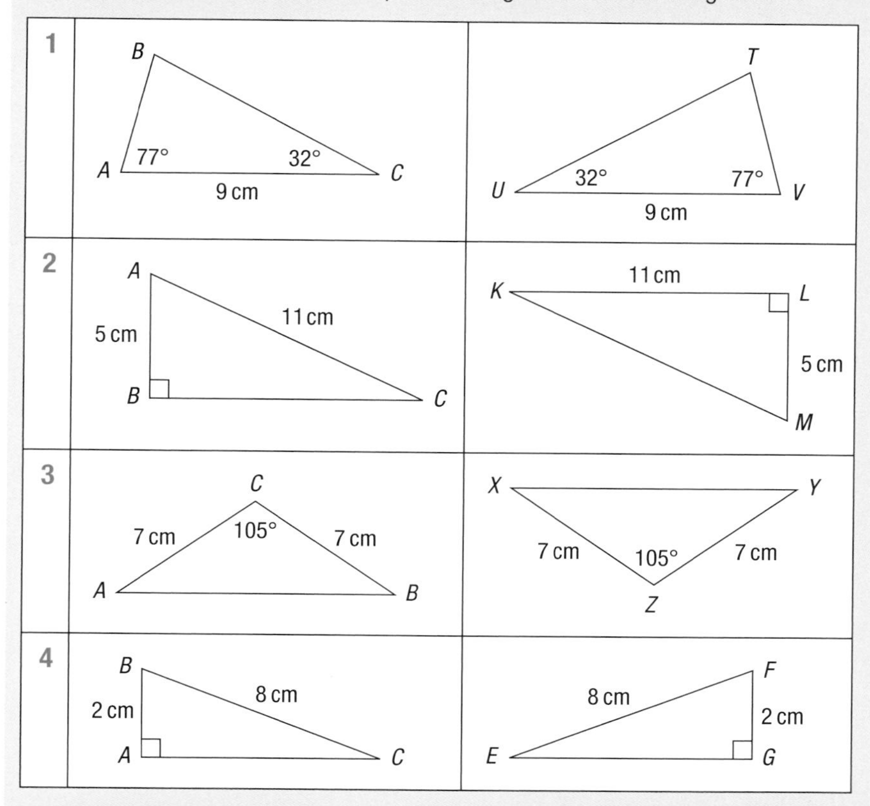

5 Annie draws a triangle with sides 5 cm, 7 cm, 8 cm. Marius draws a triangle with sides 10 cm, 14 cm, 16 cm. Annie says 'The triangles are congruent.' Explain carefully whether she is right.

Symmetry in 2D and 3D shapes

Two-dimensional (2D) shapes often have **rotation symmetry** or **reflection symmetry**. Some shapes can have both.

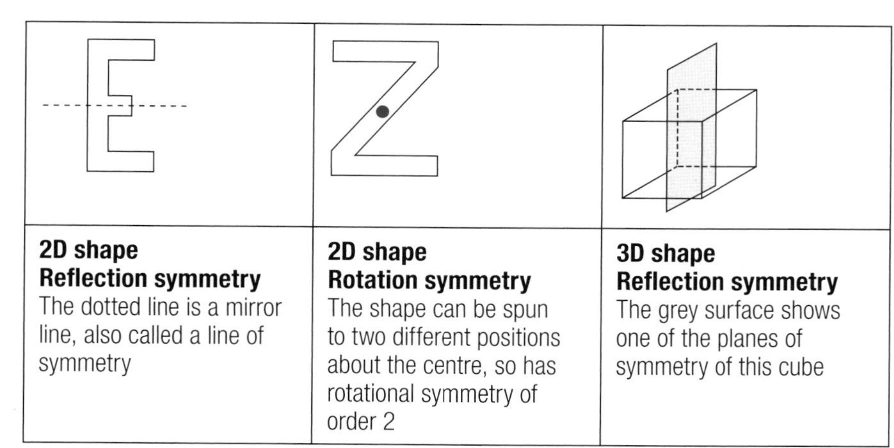

2D shape **Reflection symmetry** The dotted line is a mirror line, also called a line of symmetry	2D shape **Rotation symmetry** The shape can be spun to two different positions about the centre, so has rotational symmetry of order 2	3D shape **Reflection symmetry** The grey surface shows one of the planes of symmetry of this cube

Check your understanding 4.2

Describe the symmetry of these 2D shapes.

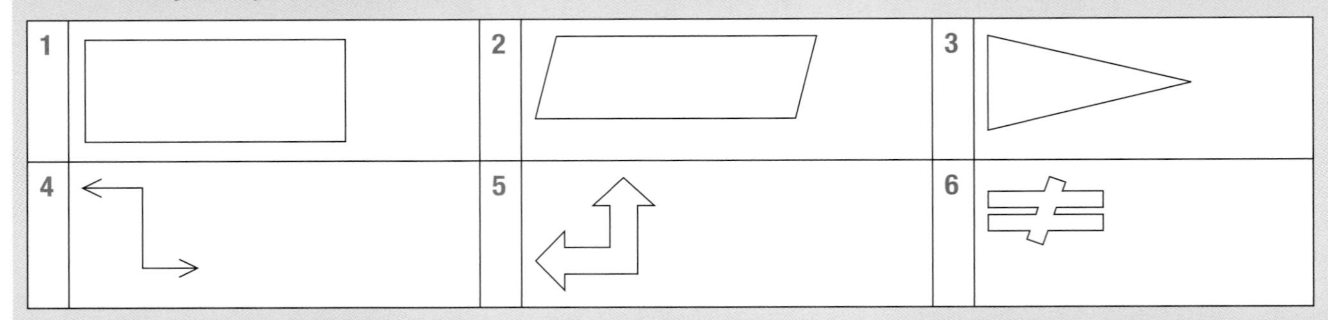

Decide whether or not each of these 3D solids has no reflection symmetry, one plane of symmetry, or more than one plane of symmetry.

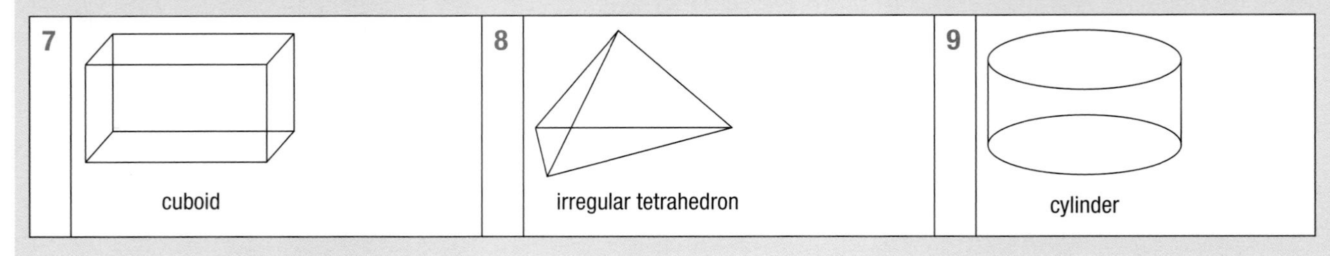

Angles in polygons

A polygon is formed by a number of straight sides. At each corner, or **vertex**, there is an **interior** angle and an **exterior** angle. The exterior angles always **add up to 360°**.

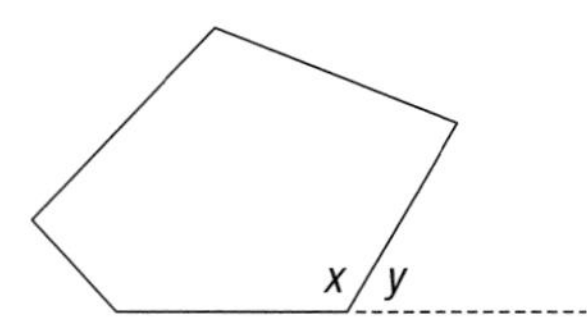

x = internal angle
y = external angle

$x + y = 180°$

Worked example

The diagram shows a regular pentagon.

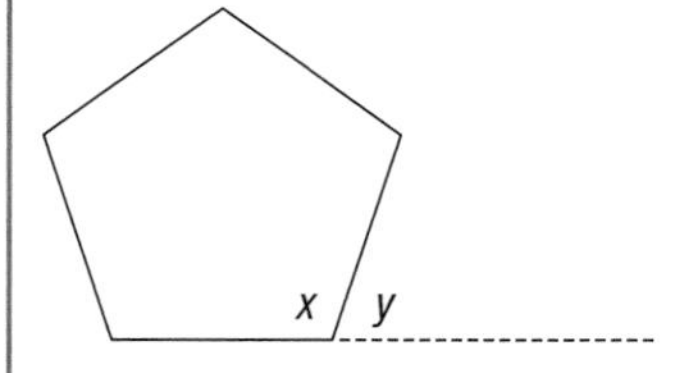

Find the angles marked x and y.

Solution
Since this is a regular pentagon, the exterior angles are all equal.
So $y = 360° \div 5 = 72°$.
The interior and exterior angles add up to 180°.
So $x = 180° - 72° = 108°$.

Check your understanding 4.3

1 a) State the number of sides of a regular hexagon.
  b) Work out the size of each external angle of a regular hexagon.
  c) Hence find the size of each internal angle of a regular hexagon
2 A regular polygon has ten sides. Find the size of each external angle.
3 A regular polygon has nine sides. Find the size of each internal angle.
4 A regular polygon has each external angle of 15°. Work out the number of sides it has.
5 A regular polygon has each internal angle of 172°. Work out the number of sides it has.

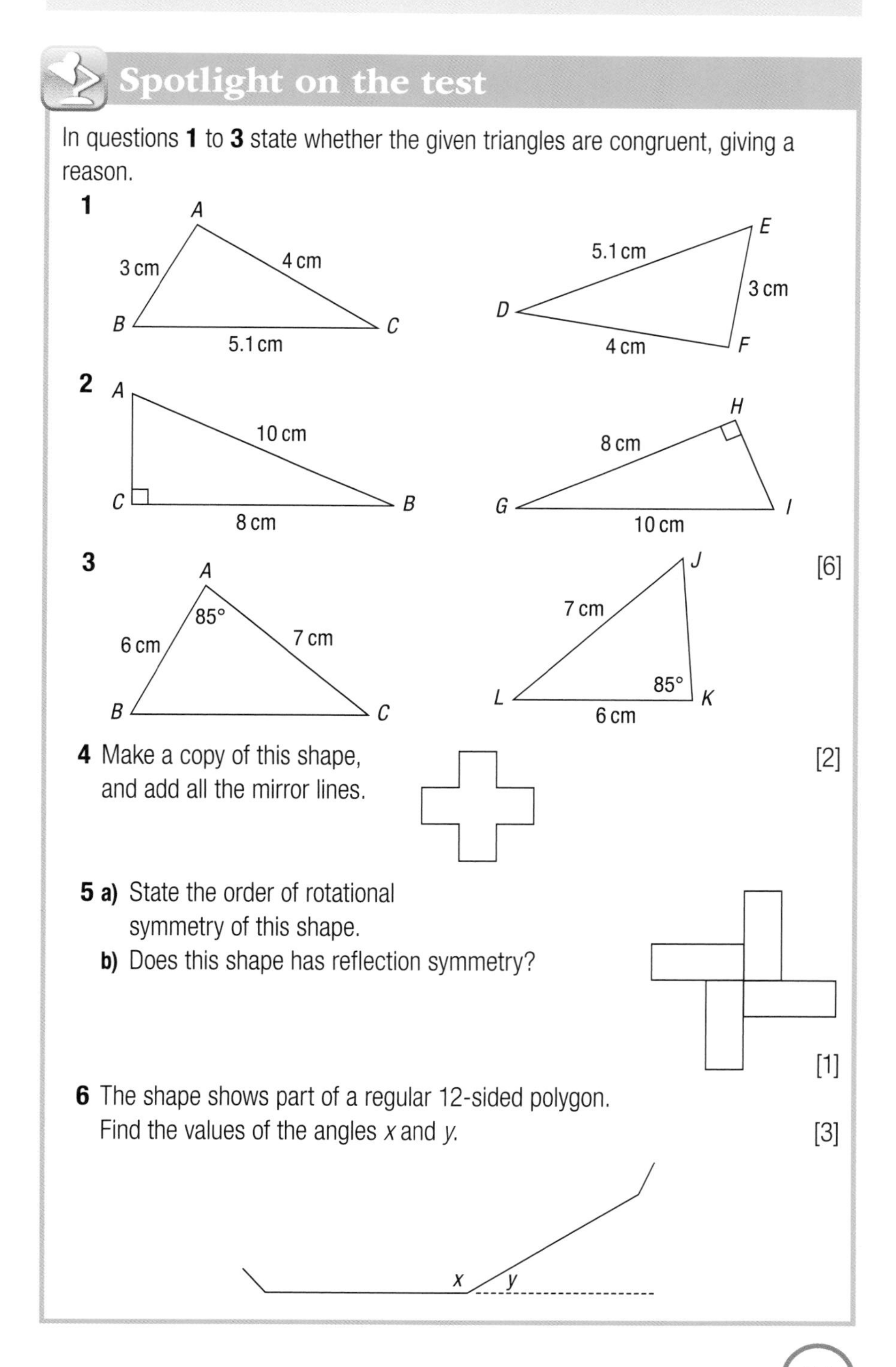

Spotlight on the test

In questions **1** to **3** state whether the given triangles are congruent, giving a reason.

1

2

3 [6]

4 Make a copy of this shape, and add all the mirror lines. [2]

5 a) State the order of rotational symmetry of this shape.
b) Does this shape has reflection symmetry? [1]

6 The shape shows part of a regular 12-sided polygon. Find the values of the angles x and y. [3]

5 Measures and motion

Metric and imperial units

Student's book references

- Book 1 Chapters 4 and 11
- Book 2 Chapter 4
- Book 3 Chapters 4 and 11

Metric units are related by powers of 10. In particular, prefixes indicate multiples of 10^3, 10^6 and so on.

Mass: 1 kilogram = 1000 grams
1 tonne = 1000 kilograms
Length: 1 kilometre = 1000 metres
1 metre = 1000 millimetres (= 100 centimetres)
Capacity: 1 litre = 1000 millilitres

Many people still use the older, imperial system of units.

Mass: Tons, stones, pounds, ounces
Length: Miles, yards, feet, inches
Capacity: Gallons, quarts, pints

You should learn that 8 kilometres is approximately 5 miles. For smaller distances, 30 centimetres (cm) is approximately 1 foot (12 inches), so 1 inch is about 2.5 cm. One kilogram is approximately 2.2 pounds.

Check your understanding 5.1

1 a) Write 1.6 metres (m) in centimetres (cm).
b) Convert 2500 millilitres (ml) into litres (l).
c) Express 3.5 kilograms (kg) in grams (g).
2 a) Write 350 cm in metres (m).
b) Convert 0.75 litre into millilitres (ml).
c) Express 650 grams in kilograms (kg).
3 The distance from London to Brighton is 45 miles. Give the equivalent distance in kilometres.
4 A pencil is 6 inches long. Convert this into centimetres.
5 The distance from Kuala Lumpur to Singapore is 304 kilometres. Roughly how many miles is this?
6 A suitcase weighs 33 pounds. Roughly how many kilograms is this?

Speed and time

Speed is an example of a **compound measure**, as it is calculated by dividing two other measures: distance by time.

Worked example

An aircraft travels 1500 km in $2\frac{1}{2}$ hours. Work out its speed in km per hour.

Solution

Speed = distance ÷ time
= 1500 ÷ 2.5
= 600 km per hour (km h^{-1})

Tips for success

- You can often work out speeds by scaling, rather than dividing distance by time. For example, travelling 12 kilometres in 30 minutes is the same rate as 24 km in 60 minutes, that is, it is a speed of 24 km per hour.
- This scaling method works particularly well when you are told how far something travels in say 10 minutes, 12 minutes, 15 minutes or 20 minutes – all of these will scale easily to 60 minutes.

Worked example

Monica can walk 800 m in 10 minutes. Dara can walk 900 m in 12 minutes. Work out their speeds, in km h^{-1}, and hence decide who the faster walker is.

Solution

1 hour is 6 × 10 minutes, that is, 6 × 800 m = 4800 m

So Monica walks at 4.8 km h^{-1}.

1 hour is 5 × 12 minutes, that is, 5 × 900 m = 4500 m

So Dara walks at 4.5 km h^{-1}.

Hence Monica is the faster walker.

Check your understanding 5.2

1. A car drives 460 km in 4 hours. Work out its speed in km per hour.
2. A sprinter runs 200 metres in 25 seconds. Work out his speed in metres per second.
3. A ferry takes 1 hour 30 minutes to travel 24 miles. Work out its speed in miles per hour.
4. In 2012, an 11-year-old cheetah named Sarah was timed in a sprint by Cincinnati Zoo in the USA. Sarah ran 100 m in 6 seconds.
   a) At this speed, how far would she run in 1 minute?
   b) Work out Sarah's speed in kilometres per hour.
5. Four snails are having a race.
   Snail A can travel 30 cm in 5 minutes.
   Snail B can travel 38 cm in 6 minutes.
   Snail C can travel 85 cm in 15 minutes.
   Snail D can travel 110 cm in 20 minutes.
   a) Work out the speed of each snail, in metres per hour.
   b) Hence work out which snail is the fastest.

Travel graphs

A travel graph shows how distance changes during a journey. Time is plotted on the x axis and distance on the y axis. The speed on any section can be worked out by looking at the steepness of the graph.

Worked examples

The graph shows a bicycle journey.

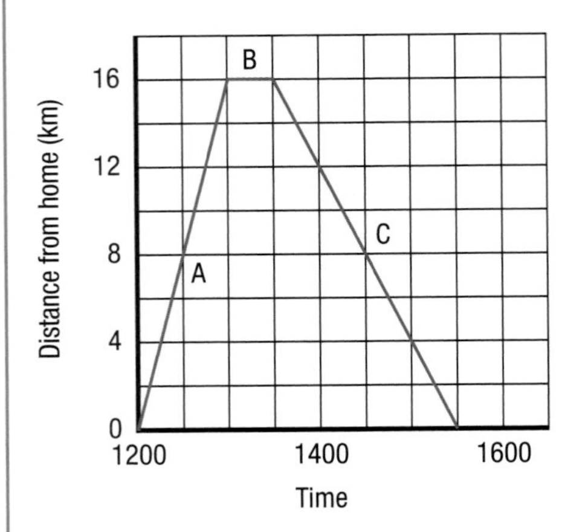

a) Work out the speed of the bicycle during section A.

b) Work out the speed of the bicycle during section C.

c) What was happening during section B?

Solution

a) During section A the bicycle travels 16 km in one hour. So its speed is 16 km per hour.

b) During section C the bicycle travels 16 km in two hours. So its speed is $16 \div 2 = 8$ km per hour.

c) During section B the bicycle was stationary.

Tips for success

- The steeper the graph, the higher the speed.
- When the graph is flat (horizontal) the motion has stopped (speed is 0).

Check your understanding 5.3

1 The diagrams show two travel graphs. Find the speed for each section marked on each graph with letters A, B, C, D, E, F, G.

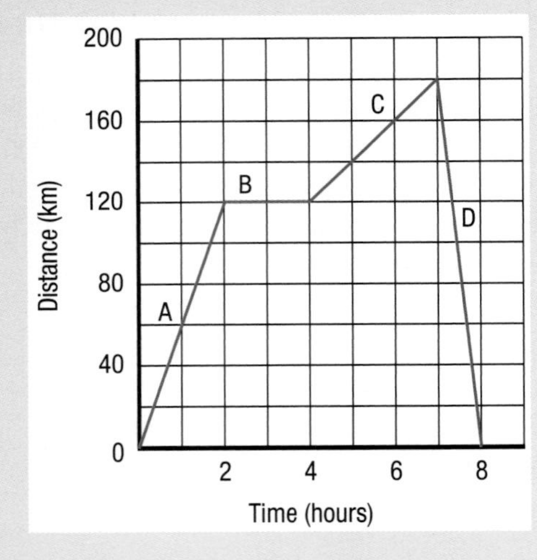

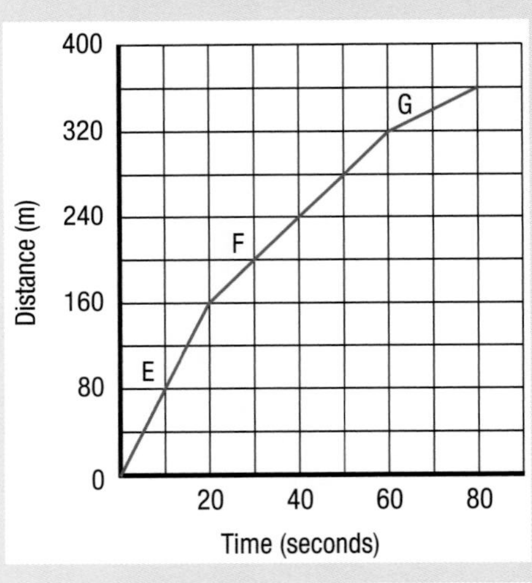

2 Misha goes to the shops. The graph shows her journey.

a) Find her speed for the journey from home to the shops.

b) Find out how long she spends at the shops.

c) Find her speed for the journey from the shops back to home.

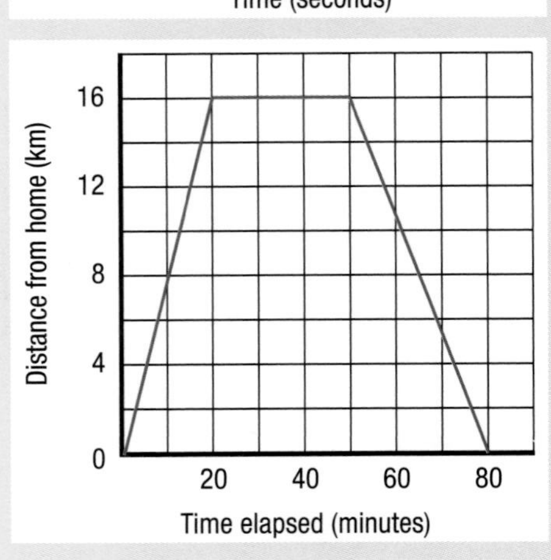

Graphs of real-life situations

Other real-life situations can also be modelled by graphs. The next activity shows you examples of how these can arise.

Tips for success

- When you meet an unfamiliar real-life graph, take a few minutes to study the graph before starting to interpret it.

Check your understanding 5.4

1 The diagram shows the amount of fuel in a minibus fuel tank at various times during a $6\frac{1}{2}$ hour journey. The minibus began the journey with a full tank.

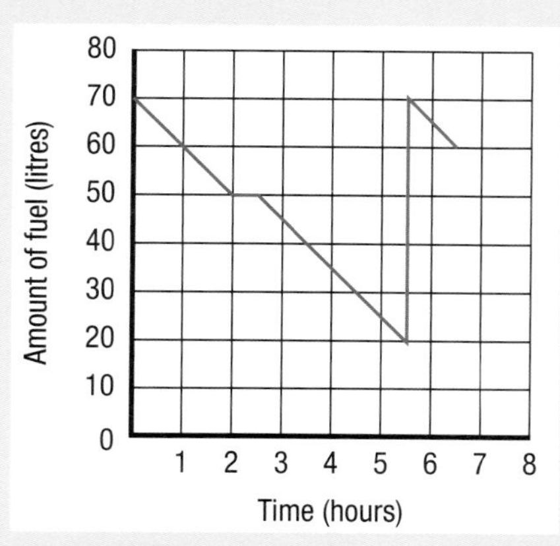

a) How much fuel can the tank hold when full?
b) How much fuel does the minibus use each hour of normal travelling?
c) Explain carefully what happened after 2 hours.
d) Explain carefully what happened after $5\frac{1}{2}$ hours.

2 In 2000, I bought a car for \$20 000 and a caravan for \$13 000. The graph shows their values at later years, until 2012.

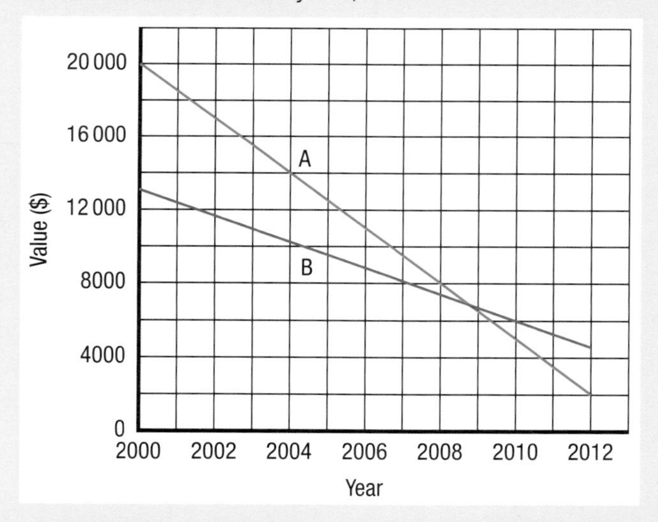

a) Which of graphs A or B shows the value of the car?
b) What was the value of the car in 2004?
c) What was the value of the caravan in 2010?
d) In which year did the value of the caravan become greater than the value of the car?

Spotlight on the test

1 Write 7500 grams in kilograms. [1]

2 Convert 3.25 metres into centimetres. [1]

3 A motorist sees a sign that the next service station is in 40 kilometres. Approximately how many miles is this? [1]

4 The travel graph shows two trains S and T.

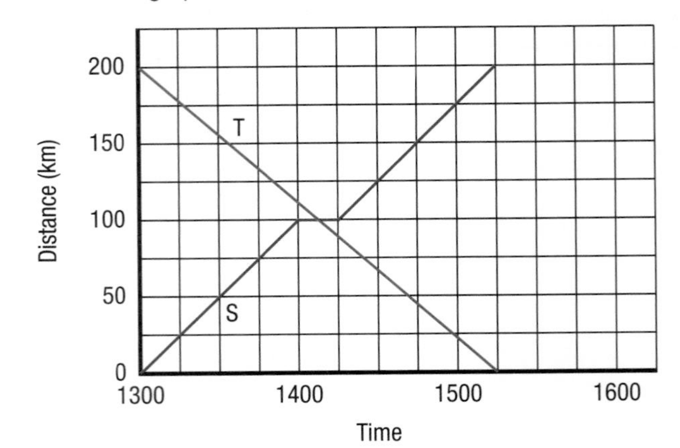

a) The trains are not supposed to travel faster than 95 km per hour. Work out whether either train is breaking this rule.

b) Explain carefully what happens at about 1408. [4]

5 The diagram below shows the height of a cable car at various times as it travels up and down a mountainside.

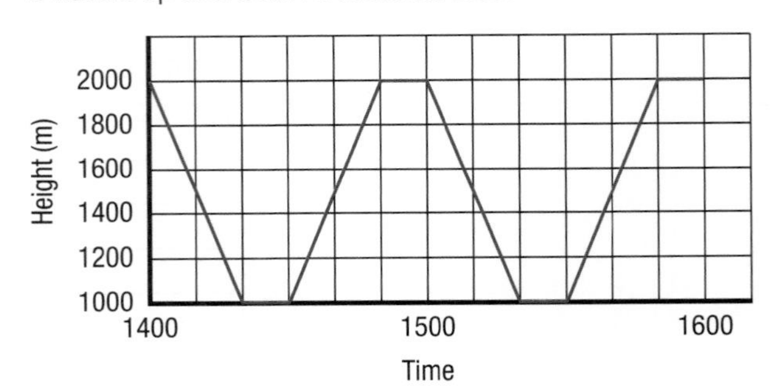

a) State whether the cable car is going up or down at 1410.

b) How long does the cable car take to travel up the mountain?

c) A passenger is waiting at the top at 1430. At what time could they expect to reach the bottom? [3]

Planning, collecting and displaying data

Discrete and continuous data

Student's book references

- Book 1 Chapter 5
- Book 2 Chapter 5
- Book 3 Chapter 5

Numerical data are usually classified as either **discrete** or **continuous**. Discrete data can only take specific values (often restricted to whole numbers) while continuous data may take any value within a given range.

Check your understanding 6.1

Decide whether each of these is discrete or continuous.

1 The number of mushrooms in a supermarket pack.
2 The time taken to complete a crossword.
3 The mass of a goldfish.
4 The number of seats on a coach.
5 The diameter of a circle.

Collecting data

Data is often collected using a **survey** and/or a **data collection sheet**. Care must be taken to make sure that the sheet allows the responder to answer each question completely and in a precise and fair way.

Check your understanding 6.2

Imagine that you are in a shopping mall on a Saturday afternoon and you are asked to fill in the following questionnaire. Decide whether you think any of the questions need to be improved.

1 How often do you come shopping to this mall?

Often fairly often sometimes rarely never

2 What gender are you?

Male female

3 How much do you expect to spend today?

\$0–10 \$10–20 \$20 or more

Stem-and-leaf diagrams

The **stem-and-leaf** diagram is a good way of organising discrete data. This enables the data to be grouped, together, yet in a way that enables all of the raw values still to be visible.

Worked example

Display these data using a stem-and-leaf diagram with stems of 150, 160, and so on: 153, 156, 157, 160, 161, 162, 162, 162, 171.

Solution

The number 153 can be written as 15|3, that is a stem of 150 and a leaf of 3. Continuing in this way we obtain this stem-and-leaf diagram:

15	3	6	7		
16	0	1	2	2	2
17	1				

Key: 15 | 3 = 153

Tips for success

- Every stem-and-leaf diagram should include a key, like the one above.
- Do not write 153 as 150|3 since this could be misread as 1503.
- Space the leaves at equal intervals, with no commas.

Check your understanding 6.3

1 The stem-and-leaf diagram shows the ages of the members of a gym club.

0	8				
1	3	5	7		
2	2	5	6	6	8
3	1	8			
4	1	3			

Key: 1 | 3 = 13

a) How many members does the club have?
b) Write down the age of the oldest member of the club.
c) What is the range of ages, between the youngest and the oldest?

2 Here are the scores obtained by 11 students in a mathematics test. The test was out of 100 marks.

71, 62, 53, 93, 65, 74, 68, 57, 74, 48, 71

Display these scores in a sorted stem-and-leaf diagram. Use stems of 40, 50, 60 and so on.

Continuous data and frequency tables

Continuous data cannot be recorded in a stem-and-leaf diagram, since the individual values cannot be measured exactly. Instead we use a frequency table, like the one below.

Length x of hand span, in cm	Frequency
$10 < x \leq 12$	2
$12 < x \leq 14$	6
$14 < x \leq 16$	7
$16 < x \leq 18$	3

Tips for success

- Notice how inequalities have been used to indicate how the boundaries between classes are to be assigned.

Data from a frequency table may be displayed in a **histogram**; this resembles a bar chart in some ways, but should have a continuous scale along the x axis. There are no gaps between the bars.

Worked example

Illustrate the data from the previous example with a histogram.

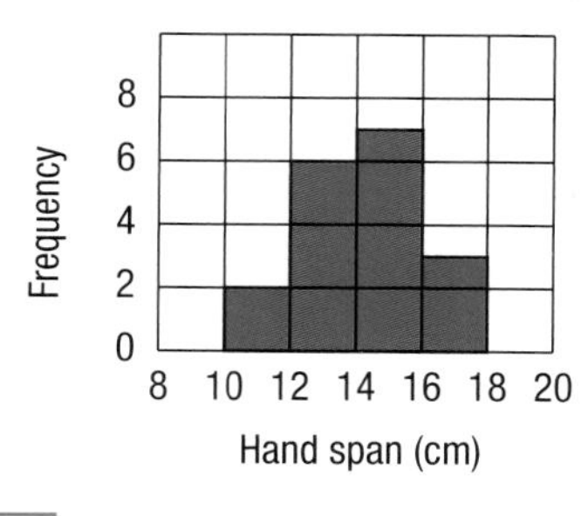

Here are three other ways of displaying data. You must be able to recognise when and when not to use each type.

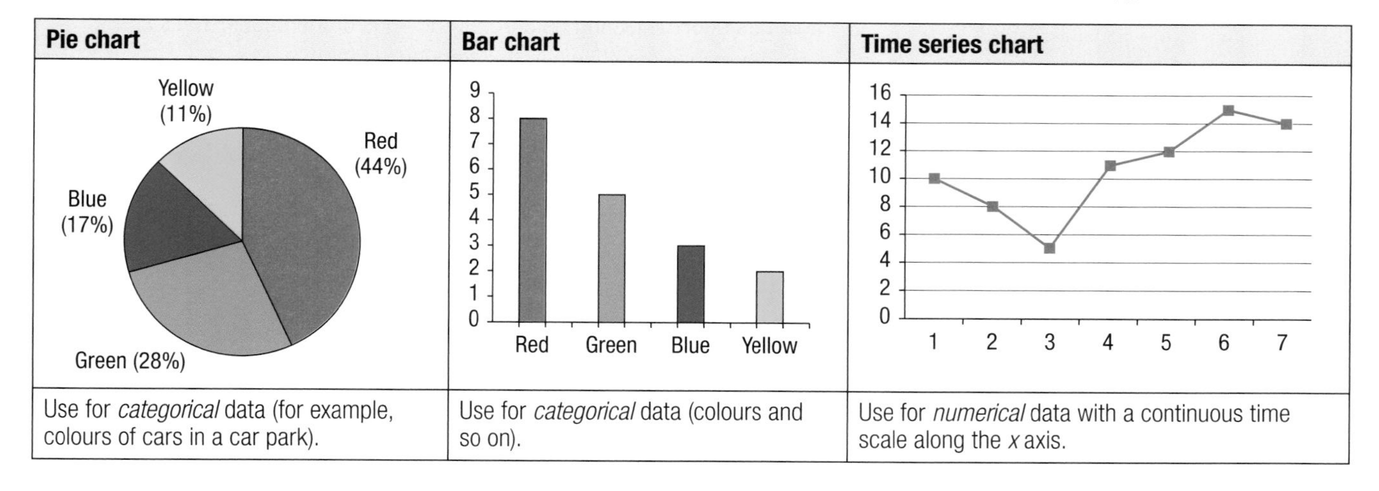

Pie chart	Bar chart	Time series chart
Use for *categorical* data (for example, colours of cars in a car park).	Use for *categorical* data (colours and so on).	Use for *numerical* data with a continuous time scale along the *x* axis.

Check your understanding 6.4

1 Here are the times of 12 runners at a school sports day. The times are given to the nearest 0.1 second.

13.8, 16.7, 12.2, 13.4, 14.0, 13.4, 13.5, 14.1, 12.5, 14.7, 13.1, 15.0

a) Copy and complete this frequency table.

Time t, in secs	Frequency
$12 \le t < 13$	
$13 \le t < 14$	
$14 \le t < 15$	
$15 \le t < 16$	
$16 \le t < 17$	

b) Illustrate the data with a histogram.

2 Here is some data from a survey about television viewing.

Favourite type of film	Frequency
Comedy	4
Horror	13
Thriller	9
Other	6

Suggest a type of diagram that would be suitable for displaying this data.

Spotlight on the test

1 Mafe is collecting data about mobile phone texting. Here is one of his questions: How many SMSs do you send per day?

1–5	5–10	10–15	15 or more
☐	☐	☐	☐

Suggest *two* ways in which this question could be improved. [2]

2 David counts the number of passengers on some coaches. Here are the results:

24, 12, 19, 15, 13, 52, 36, 18, 21, 27, 31, 13

a) Display these results in a stem-and-leaf diagram, with stems of 10, 20, 30. Remember to include a key.

b) One value is very different from all the others. Say which one it is. [4]

3 Paulo has been collecting data about the ages of the people in his sports club. The results are shown in this frequency table.

Age n of child, in years	Frequency
$5 \le n < 10$	3
$10 \le n < 15$	6
$15 \le n < 20$	9
$20 \le n < 25$	8
$25 \le n < 30$	5

After Paulo collected this data, two new members joined the club. Their ages are 9 and 10.

a) Draw up a new frequency table to include all the members of the club.

b) Illustrate the new table with a histogram.

c) Explain briefly why a pie chart would not be suitable. [5]

Equations, functions and inequalities

Linear equations

Student's book references

- Book 1 Chapter 9
- Book 2 Chapter 9
- Book 3 Chapter 9

Tips for success

- You can add, subtract, multiply or divide one side of an equation by a number, as long as you do the same thing to the other side.
- Record the steps in brackets, as in the worked examples here.

A **linear equation** is one like $5x + 3 = 23$. Such equations can be **solved** by gathering all the x terms on one side of the equation, and the number terms on the other.

Worked example

Solve the equation $5x + 3 = 23$.

Solution

$5x + 3 = 23$

$5x = 23 - 3$ (subtracting 3 from each side)

$5x = 20$

$x = 20 \div 5$ (dividing both sides by 5)

$x = 4$.

Sometimes linear equations require clearing brackets first.

Worked example

Solve the equation $6(x + 2) = 36 - 2x$.

Solution

$6(x + 2) = 36 - 2x$

$6x + 12 = 36 - 2x$ (clearing the brackets)

$6x + 12 + 2x = 36$ (adding $2x$ to both sides)

$8x + 12 = 36$

$8x = 36 - 12$ (subtracting 12 from both sides)

$8x = 24$

$x = 24 \div 8$ (dividing both sides by 8)

$x = 3$.

Check your understanding 7.1

Solve these linear equations.

1. $x + 5 = 12$
2. $2x = 16$
3. $4 + x = 21$
4. $4x = 2$
5. $\frac{x}{5} = 2$
6. $5x = 2x + 9$
7. $10 + x = 12 - x$
8. $10 - 2x = 0$
9. $3(x + 1) = 6$
10. $5(2x - 1) = 35$
11. $5x + 3 = 27 - x$
12. $2x + 16 = 1 + 7x$
13. $4(2x - 1) = 26 + 2x$
14. $5x + 4 = 60 - 2x$
15. $13x + 1 = 3(x + 7)$
16. $5(2x + 3) = 7(x + 3)$

Simultaneous equations

Sometimes you meet two linear equations, each containing two variables, x and y say. These can often be **solved simultaneously**, that is we seek one set of x and y values that satisfy both equations at the same time. A common method is algebraic elimination.

Worked example

Solve the equations:

$$9x + 5y = 13$$
$$3x + 2y = 4$$

Solution

Note that $5y$ and $2y$ can both be scaled up to $10y$. So, we have

$$18x + 10y = 26 \text{ (first equation } \times 2)$$
$$15x + 10y = 20 \text{ (second equation } \times 5)$$

Subtracting,

$$3x\ (+\ 0y) = 6$$

So, dividing both sides by 3,

$$x = 2$$

Now substitute this value into the original first equation,

$$9x + 5y = 13$$
$$9 \times 2 + 5y = 13$$
$$18 + 5y = 13$$
$$5y = 13 - 18$$
$$5y = -5.$$

So, dividing both sides by 5,

$$y = -1.$$

The full solution is $x = 2$ and $y = -1$.

In the above example, the $10y$ terms were identical (both positive) so we subtracted one equation from the other; this would also be the method if they were both negative. However, when one is positive and the other is negative you have to add the two equations, as in this next example.

Worked example

Solve the equations:

$$3x + 4y = 11$$
$$7x - 3y = 1$$

Solution

$$3x + 4y = 11$$
$$7x - 3y = 1$$

$9x + 12y = 33$
$28x - 12y = 4$
$37x = 37$
$x = 1.$

Then substituting back into the original first equation:

$3 \times 1 + 4y = 11$
$3 + 4y = 11$
$4y = 11 - 3$
$4y = 8$
$y = 2.$

So the solution is $x = 1$ and $y = 2$.

Tips for success

- If the equal matching terms to be eliminated are the same sign – both positive or both negative – then you *subtract* … **S**ame, **S**ubtract.
- If the matching terms are equal but different then you *add* them instead.

Check your understanding 7.2

Solve these pairs of simultaneous equations, using the elimination method.

1 $6x + 2y = 20$
$2x - y = 5$

2 $4x - 2y = 12$
$5x + y = 22$

3 $3x + y = 21$
$5x + 2y = 36$

4 $7x - 2y = 23$
$3x - 2y = 11$

5 $2x - y = 11$
$x + y = 10$

6 $7x + 2y = 50$
$2x + y = 16$

7 $4x - 3y = 11$
$x - 2y = 4$

8 $11x + 5y = 16$
$3x + 2y = 5$

9 $5x - 2y = 19$
$4x + 3y = 6$

10 $4x + 3y = 28$
$5x + y = 35$

Constructing and solving equations

Some questions will ask you to form an equation from a real-world situation, then go on to solve it.

Worked examples

A rectangle has length $2x + 1$ cm and width x cm as shown in the diagram.

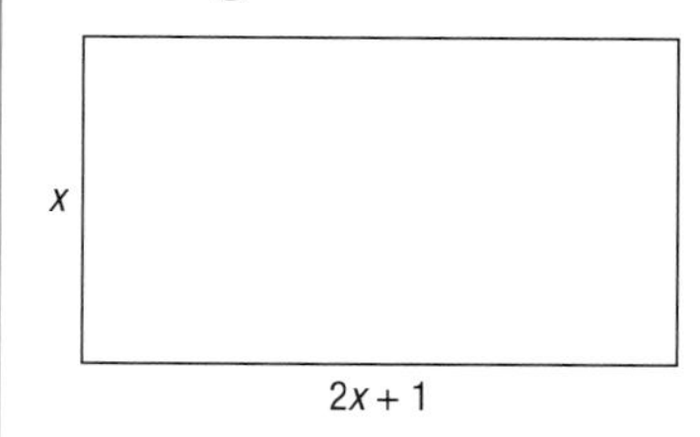

The perimeter of the rectangle is 44 cm.

a) Write this information as an equation.

b) Solve your equation, to find the value of x.

c) Hence state the dimensions of the rectangle.

Solution

a) Adding the four sides of the rectangle,

$$(2x + 1) + x + (2x + 1) + x = 44.$$

b) Simplifying,

$$2x + 1 + x + 2x + 1 + x = 44$$
$$6x + 2 = 44$$
$$6x = 42.$$

So, dividing by 6,

$$x = 7.$$

c) Then the dimensions of the rectangle are $2x + 1$ and x. So the dimensions of the rectangle are 15 cm and 7 cm.

Check your understanding 7.3

1 The diagram below shows a rectangle.

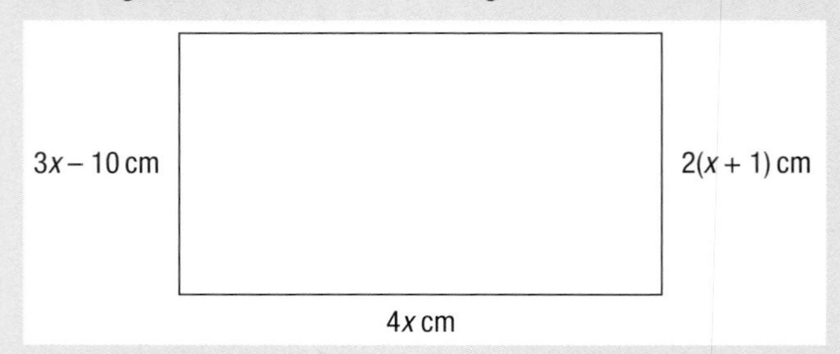

a) Write down an equation in x.

b) Solve your equation, to find the value of x.

c) Hence find the dimensions of the rectangle.

2 A bus can carry 40 passengers. It begins its journey with y passengers on board. At the first stop another y passengers get on, and six get off. At the next stop $2y$ passengers get on and two get off; the bus is now full.

a) Write this information as an equation in y.

b) Solve your equation, to find the value of y.

3 The three sides of a triangle have lengths of $3x - 2$, $4x + 2$ and $x + 7$ cm. The perimeter of the triangle is 39 cm.

a) Write down an equation in x. (You do not need to simplify it yet.)

b) Work out the value of x.

c) Now find the lengths of each of the sides of the triangle.

4 Some teenagers are going on a minibus trip. The minibus charges a fixed $10 booking fee, plus $12 for each person on the trip. The total cost is $$C$ when n people go on the trip.

a) Write this information as a formula in the form $C = \dots$.

b) In fact the total cost turned out to be $118. Use algebra to work out the number of people on the trip.

Linear inequalities

Linear inequalities work in a similar way to equations. You can add or subtract the same quantity to or from both sides, or you can multiply or divide both sides by the same positive quantity.

Worked example

Solve the inequality $2x + 5 < 13$.

Solution

$2x + 5 < 13$

$2x < 13 - 5$ (subtract 5 from both sides)

$2x < 8$

$x < 8 \div 2$ (dividing both sides by 2)

$x < 4$.

Inequalities may be illustrated using a number line.

Worked example

Illustrate the inequality $3 \leq x < 10$ on a number line.

Solution

The value x must lie in the region between 3 and 10, including 3 but not 10:

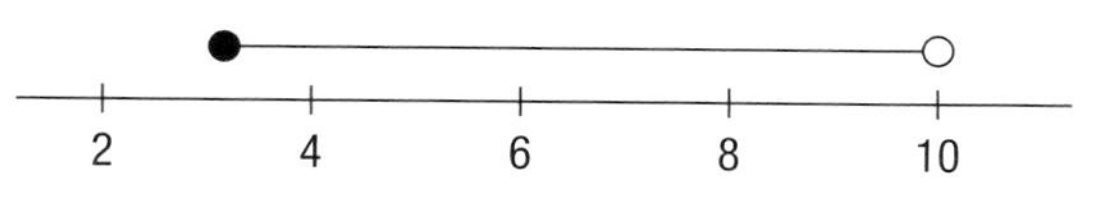

Tips for success

- Notice how the use of the bulbs at the end of the line allows you to show that 3 is included but 10 is not.

Check your understanding 7.4

Solve these inequalities.

1 $3x - 1 < 11$

2 $2x + 9 \leq 21$

3 $5y + 13 < 38$

4 $x + 7 \leq 15 - x$

5 $3x + 2 \leq 2x + 3$

6 $6 < 10 - x$

7 $11 + x < 31 - 3x$

8 $2(x + 3) \leq 20$

Illustrate these inequalities on a number line.

9 $5 < x \leq 11$

10 $3 \leq x < 7$

11 $4 \leq 2x \leq 16$

12 $8 < 4x < 24$

Find all the whole numbers (integers) that satisfy each set of conditions.

13 $2 < m \leq 7$

14 $3 \leq n < 8$

15 $4 \leq 2m \leq 18$

16 $18 < n < 20$

Multiplying out a pair of brackets

In Chapter 3 you revised how to multiply a bracket by a single term. You can also multiply two brackets together. The grid method is quite helpful for this.

Worked example

Expand and simplify: $(2y + 7)(3y - 5)$.

Tips for success

- You may have used a method called FOIL, or 'smiles and eyebrows', for expanding two brackets, but the grid method is a better way of organising the results, and is better suited to harder problems to be studied later.

Solution

$\times$	$2y$	$+7$
$3y$	$6y^2$	$+21y$
-5	$-10y$	-35

Using the table, we see that:

$$(2y + 7)(3y - 5) = 6y^2 + 21y - 10y - 35$$
$$= 6y^2 + 11y - 35$$

Check your understanding 7.5

Expand these brackets, simplifying the results.

1 $(x + 3)(x + 4)$
2 $(y + 6)(y + 2)$
3 $(x - 7)(x + 2)$
4 $(y + 5)(y - 4)$
5 $(x - 11)(x + 2)$
6 $(x + 6)(x - 6)$
7 $(y - 3)(y + 3)$
8 $(y + 3)(y + 2)$
9 $(2x - 1)(x + 1)$
10 $(2x + 3)(x - 2)$
11 $(3x - 2)(x - 1)$
12 $(3x - 1)(x + 3)$
13 $(2x + 1)(2x + 3)$
14 $(3x + 1)(2x - 3)$
15 $(3x + 2)(2x + 3)$
16 $(3x + 2)(3x - 2)$

Spotlight on the test

1 You are given that $7x - 1 = 23 - x$. Use algebra to find the value of x. [3]

2 Solve the simultaneous equations

$$4x + 2y = 22$$
$$3x - y = 14$$ [4]

3 Each side of a regular pentagon has a length $2y + 6$. Find an expression for its perimeter. Give your answer in simplified form. [2]

4 Shape A is an equilateral triangle. Shape B is a square. The **perimeters** of shapes A and B are equal. Work out the lengths of the sides of each shape, using algebra.

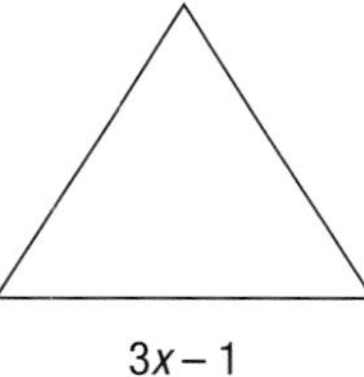

$3x - 1$

SHAPE A

$x + 3$

SHAPE B [3]

5 Find all the positive whole numbers n that satisfy $3n - 1 < 10$. [2]

6 a) Solve the inequality $12 - x < x \leq 20 - x$.

b) Illustrate this solution on a number line. [2]

7 Expand and simplify the bracketed expression: $(x + 5)(x - 3)$. [2]

8 Expand and simplify the bracketed expression: $(2x + 3)(3x - 1)$. [3]

Measurement and construction

Geometric constructions

Student's book references

- Book 1 Chapter 10
- Book 2 Chapter 10

Here is a reminder of three important geometric constructions.

A triangle with three sides given (SSS)	Perpendicular bisector of a line segment	Bisector of a given angle
1 Start with…drawing the longest side	**1** Start with…a given line segment	**1** Start with…a given angle
2 Use compasses for the middle side	**2** Construct equal arcs above the line	**2** Construct equal arcs from the angle
3 Compasses for the short side	**3** Construct equal arcs below the line	**3** Construct new arcs centred on the old ones
4 Join to complete the triangle	**5** Join to complete	**4** Join to complete

Check your understanding 8.1

1. Use compasses and a ruler to construct a triangle with sides of lengths 4 cm, 5 cm and 7 cm.
2. Use compasses and a ruler to construct a triangle with sides of lengths 5 cm, 6 cm and 9 cm. Now use your compasses to bisect the largest angle.
3. Draw a line segment of length 8 cm. Use your compasses and a straight edge to construct the perpendicular bisector of this line segment.

Spotlight on the test

1 Construct an equilateral triangle with sides of length 5 cm. Leave all your construction lines visible. [2]

2 Construct a triangle with sides 3 cm, 4 cm and 5 cm. Now bisect the largest angle. [4]

3 Construct a triangle with sides of 4 cm, 6 cm and 7 cm. Now construct the perpendicular bisector of the longest side. [4]

4 Construct a regular hexagon with sides of length 5 cm. Leave all your construction lines visible. [3]

5 The diagram below shows how a student has constructed a line *BF* that divides the angle *ABC* into two equal parts.

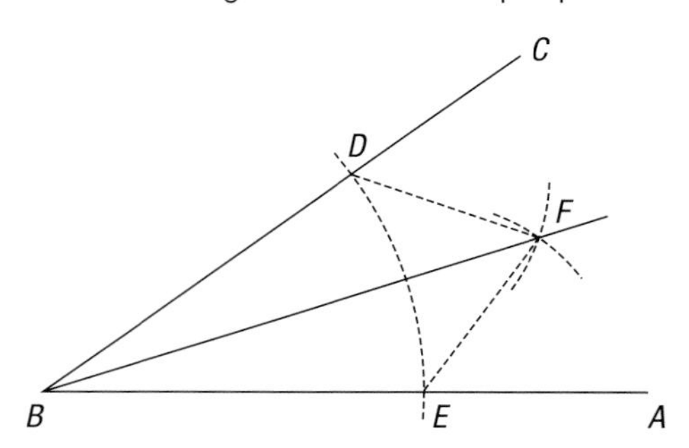

The student carried out a set of instructions. Here are the instructions, in the *wrong* order.

- Use a ruler to draw a straight line passing through *B* and *F*.
- Use compasses to draw the arc *DE* centred on *B*.
- Draw two straight lines *BA* and *BC* meeting at a point *B*.
- The two arcs centred on *D* and *E* cross each other (intersect) at *F*.
- Use compasses to draw two arcs (with the same radii) centred on *D* and *E*.

Rewrite the instructions in the *correct* order. [2]

9 Pythagoras' theorem

Finding a hypotenuse using Pythagoras' theorem

Student's book reference

- Book 3 Chapter 10

The **hypotenuse** is the longest side in a right-angled triangle, and is always opposite to the right angle.

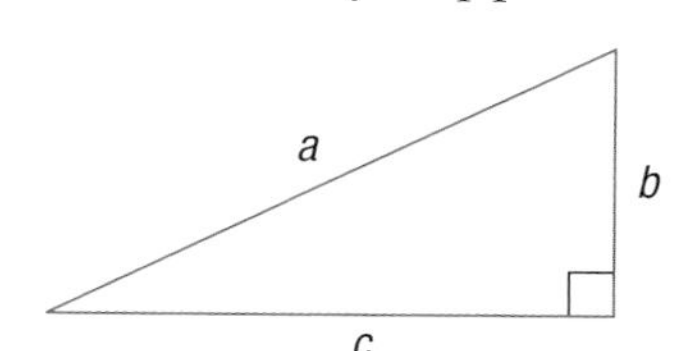

Here a is the hypotenuse, and b and c are the two other, shorter, sides. Pythagoras' theorem tells us that:

$$a^2 = b^2 + c^2$$

Worked example

Find the length x. Give your answer correct to three significant figures.

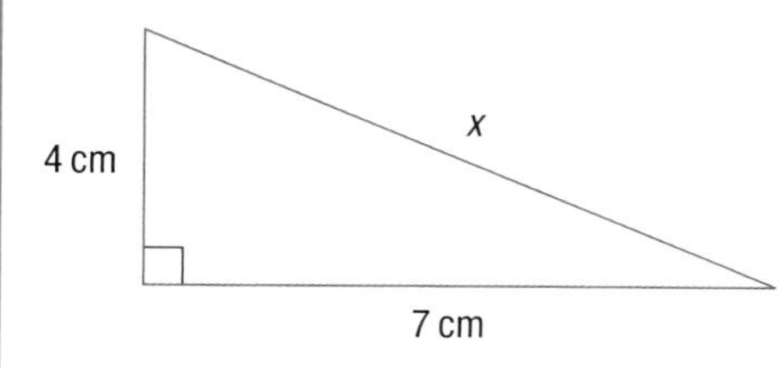

Solution

By Pythagoras' theorem:

$$x^2 = 4^2 + 7^2$$
$$= 16 + 49$$
$$= 65$$
$$x = \sqrt{65}$$
$$= 8.06\text{ cm (three significant figures)}$$

Check your understanding 9.1

Find the hypotenuse, a, in each case, given these values of b and c.

1. $b = 5$ cm, $c = 12$ cm
2. $b = 15$ cm, $c = 8$ cm
3. $b = 3.5$ cm, $c = 12$ cm
4. $b = 1.5$ cm, $c = 2$ cm
5. $b = 7$ cm, $c = 9$ cm
6. $b = 8$ cm, $c = 3$ cm
7. $b = 7$ cm, $c = 7$ cm
8. $b = 9$ cm, $c = 40$ cm

Finding a shorter side using Pythagoras' theorem

Sometimes the hypotenuse is one of the two given sides, and we have to find one of the shorter sides instead. This always involves subtraction.

Worked example

Find the length y. Give your answer correct to three significant figures.

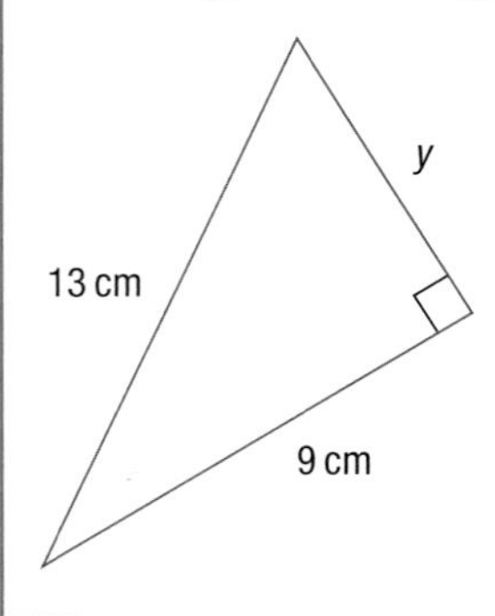

Solution

By Pythagoras' theorem:

$$13^2 = y^2 + 9^2$$

So

$$y^2 = 13^2 - 9^2$$
$$= 169 - 81$$
$$= 88$$
$$x = \sqrt{88}$$
$$= 9.38\text{ cm (three significant figures)}$$

Tips for success

- As the hypotenuse is the longest side, you always square and add for this.
- If, instead, the hypotenuse is given, then you must square and subtract.
- Show all the steps in your working, including the exact root, in case you make a slip with the final calculator stage.

Check your understanding 9.2

Find the shorter side, b, given these values of the hypotenuse a and the third side c.

1 $a = 2.5$ cm, $c = 2$ cm
2 $a = 8.5$ cm, $c = 4$ cm
3 $a = 10$ cm, $c = 6$ cm
4 $a = 20.5$ cm, $c = 4.5$ cm
5 $a = 8$ cm, $c = 5$ cm
6 $a = 11$ cm, $c = 7.5$ cm
7 $a = 16$ cm, $c = 15$ cm
8 $a = 7.5$ cm, $c = 4.5$ cm

Spotlight on the test

1 Find the lengths represented by letters. [8]

a
4 cm
13 cm

3 cm
6 cm
b

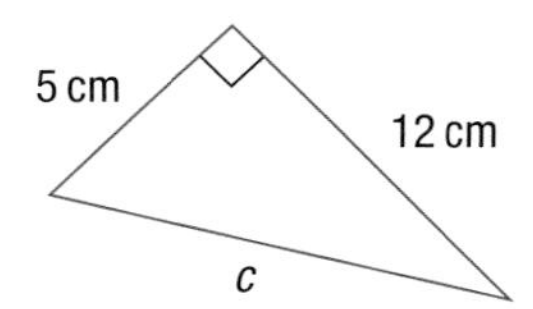

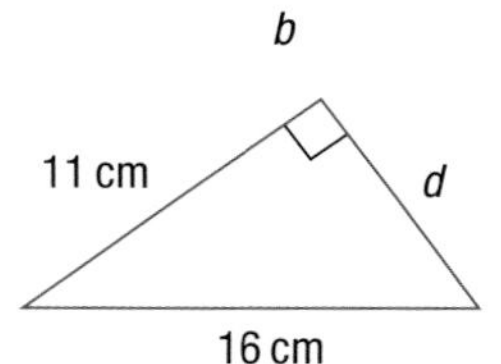

2 Find the lengths marked j and k in the diagram below. [4]

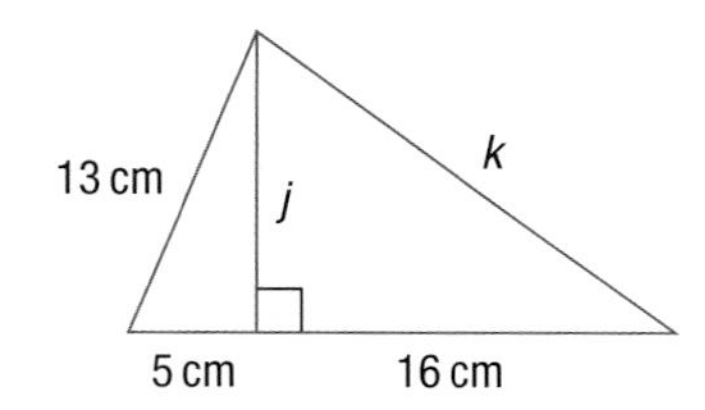

10 Transformations

Enlargement

Student's book references

- Book 1 Chapter 1
- Book 2 Chapters 11 and 19
- Book 3 Chapters 12 and 19

An **object** can be enlarged by a given scale factor to form an **image**.

Worked example

The diagram shows a rectangle **R**. Enlarge **R** by scale factor 2, centre (0,2).

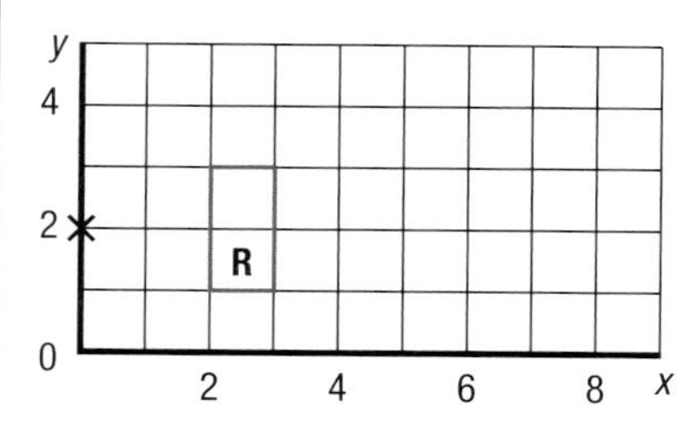

Solution

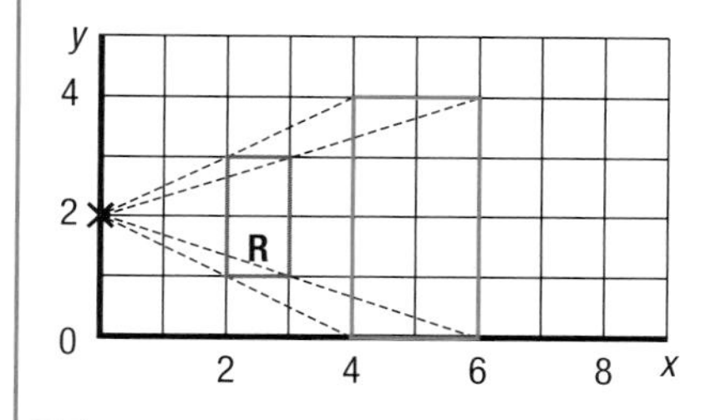

Check your understanding 10.1

1 The diagram shows an object **A** and its image **B** after an enlargement. State the scale factor for the enlargement, and give the coordinates of the centre of enlargement.

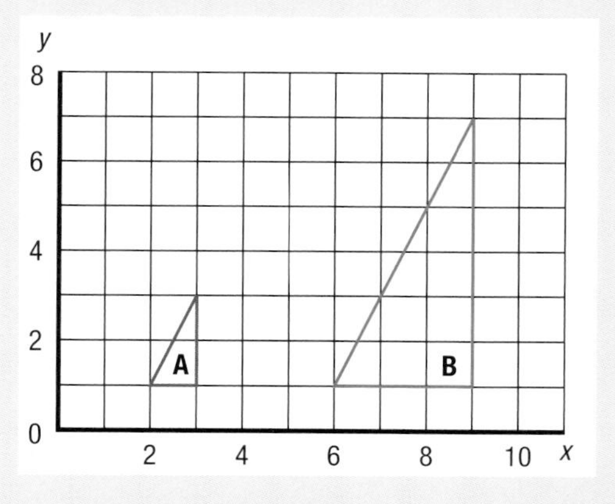

2 Triangle A has sides of length 3, 4 and 5 cm. Its enlargement, triangle **B**, has sides of length 12, 16 and 20 cm. State the scale factor for this enlargement.

Rotation, reflection, translation

Here are some examples of **rotation**, **reflection** and **translation**. Note that the object and image are **congruent** in each case.

Translation	Reflection	Rotation
The **object** slides across and up to form the **image**.	The **object** reflects in the mirror line to form the **image**.	The **object** rotates about a fixed point to form the **image**.
This is a translation of 5 units right and 1 up, which can be written as a column vector $\begin{bmatrix} 5 \\ 1 \end{bmatrix}$.	This is a reflection in the line $x = 2$.	This is a rotation of 90° clockwise about the origin.

Worked examples

The diagram shows a triangle marked **S**.

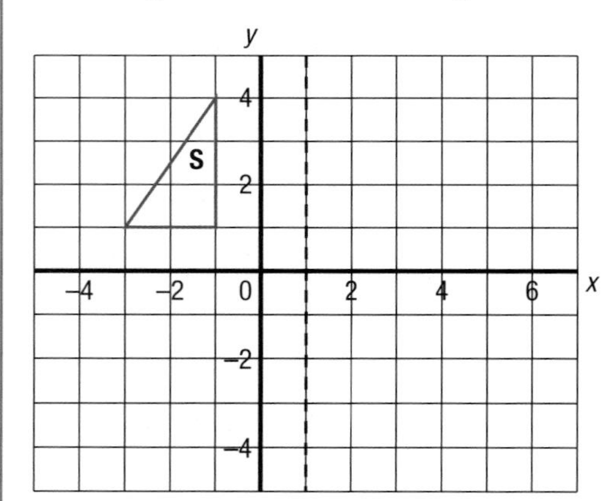

a) Reflect **S** in the line $x = 1$ to form triangle **T**.

b) Reflect **T** in the x axis to form triangle **U**.

c) Describe the transformation that would take **S** directly to **U**.

Tips for success

- When describing a translation, use the column vector form.
- For a reflection, always state the location of the mirror line.
- For rotations, include the angle, direction (unless it is 180°) and the coordinates of the centre of rotation.

Solution

a), b)

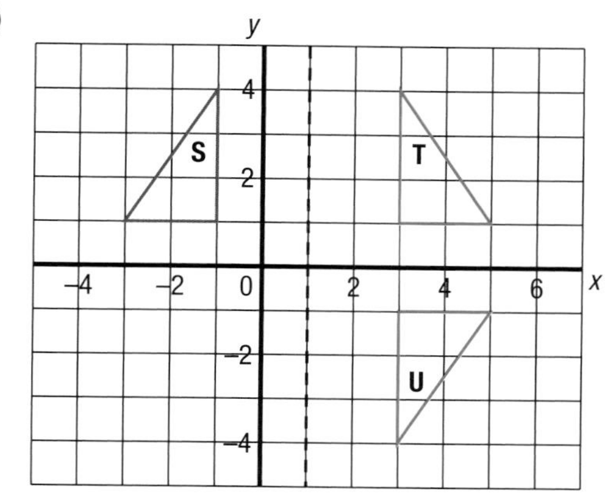

c) **S** can be transformed to **U** by a 180° rotation about (1, 0).

Check your understanding 10.2

1 The diagram shows a triangle **T**. The images of **T** under three different transformations are shown, and labelled **P**, **Q** and **R**.

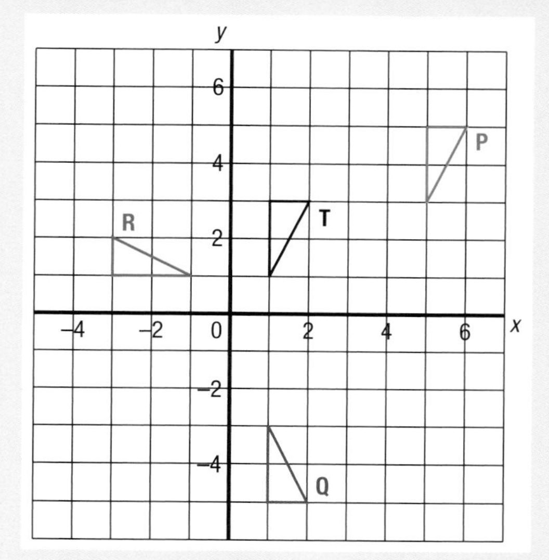

a) State the transformation that takes **T** to **P**.
b) State the transformation that takes **T** to **Q**.
c) State the transformation that takes **T** to **R**.

2 A triangle **S** is reflected in the x axis to make an image **U**. Then **U** is reflected in the y axis to make **V**. Describe the single transformation that would take **S** directly to **V**.

Spotlight on the test

1 Make a copy of this grid, showing the rectangle **R**.

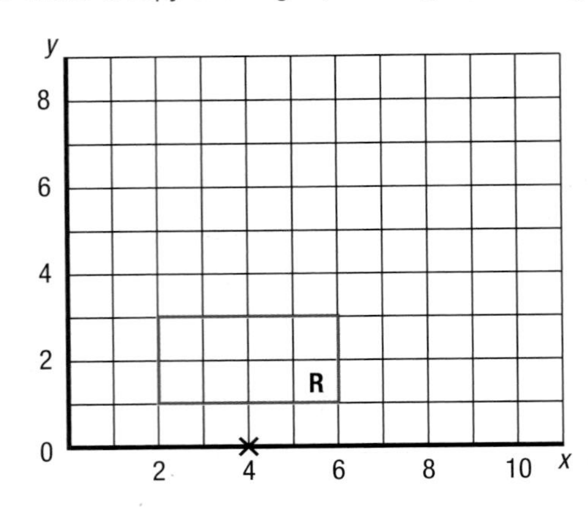

S is the enlargement of **R** with scale factor ×2, centre of enlargement (4, 0).
Draw and label **S** on your diagram. [2]

2 Make a copy of this grid, showing the triangle **T**.

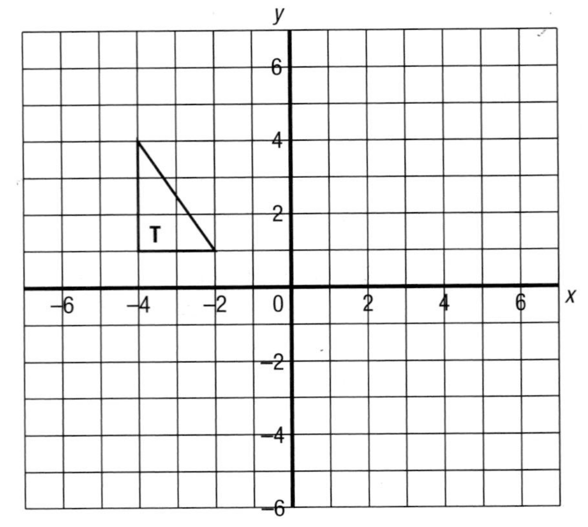

a) Translate triangle **T** by $\begin{bmatrix} 8 \\ 2 \end{bmatrix}$. Label the result **A**.

b) Rotate triangle **T** through 180° about (1, 0). Label the result **B**.

c) Describe the single transformation that would take **A** directly to **B**. [5]

3 Triangle **A** is enlarged by scale factor ×2, centre (0,3) to form triangle **B**.
Then **B** is enlarged by scale factor ×3, centre (0, 3) to form triangle **C**.
Describe the single transformation that takes **A** to **C**. [3]

11 Averages and spread

Mean, median, mode, range

Student's book references

- Book 1 Chapter 1
- Book 2 Chapters 12 and 19
- Book 3 Chapters 12 and 19

There are several different ways of describing the **average** of a set of numerical data:

- The **mean** is the simple average. Add the numbers up, and divide by the total number of items.
- The **median** is the middle value, once the data are arranged in order.
- The **mode** is the most commonly occurring value.
- You can also measure the **spread** of data in a number of ways, including the **range**, which is the difference between the highest and lowest values.

Worked example

Here are the temperatures of nine cities at noon one day in winter:

12 3 9 16 7 12 2 15 8

Find the mean, median and mode and range of the temperatures.

Solution

Arrange the values in order:

2 3 7 8 9 12 12 15 16

The total is $2 + 3 + \ldots + 16 = 84$.

The mean is $84 \div 9 = 9.3$ (to one decimal place).

The mode is 12 (most commonly occurring value).

The median is 9 (since there are four values below this and four above).

The range is $16 - 2 = 14$.

Tips for success

- A quick way to locate the median is to count the number of items (9) and add 1 (10). Then divide by two (5) and this tells you to look at the fifth number. If you had an even number of items, say 14, this method would suggest you look at the 7.5th number – so you would take the seventh and eighth numbers and find their simple average.

Check your understanding 11.1

1 In five mathematics tests Desi scored 12, 19, 15, 14 and 18.
  a) Find the mean score.
  b) Find the range of the scores.

2 Here are the noon temperatures of 12 cities, in °C:

  13, 15, 16, 17, 17, 17, 19, 22, 22, 23, 25, 28

  Find the mean, median, mode and range of this data.

3 In seven cricket matches I scored 0, 15, 22, 5, 0, 63 and 35 runs.
  a) Find the median score.
  b) Find the mode of the scores.
  c) Which one of these gives a better idea of my 'average' score?

Mean from a frequency table

Sometimes you need to find the mean from a frequency table. If the data are exact (discrete) you can calculate the mean precisely.

Worked example

The table shows the number of sweets in each of 20 tubes.

Number of sweets, n	Frequency, f
9	5
10	7
11	6
12	2

Work out the mean number of sweets in a tube.

Solution

Number of sweets, n	Frequency, f	$n \times f$
9	5	45
10	7	70
11	6	66
12	2	24
Total	**20**	**205**

Mean = 205 ÷ 20 = 10.25.

If the data have been grouped then you must work with the **mid-point** of each class. Some precision is lost: you obtain an **estimate** of the mean. You might also need to find the **modal class**: this is the class with the highest frequency.

Worked examples

The table shows the times taken by 15 workers to travel to work each day.

a) Calculate an estimate of the mean travelling time.

b) State the modal class.

Time, t (minutes)	Frequency, f
$0 < t \leq 10$	2
$10 < t \leq 20$	8
$20 < t \leq 30$	3
$30 < t \leq 40$	2

Solution

a)

Time, t (minutes)	Frequency, f	Mid-point, M	$M \times f$
$0 < t \leq 10$	2	5	10
$10 < t \leq 20$	8	15	120
$20 < t \leq 30$	3	25	75
$30 < t \leq 40$	2	35	70
Total	**15**		**275**

Mean = 275 ÷ 15 = 18.3 (one decimal place).

b) The modal class is $10 < t \leq 20$ as this has the highest frequency.

Tips for success

- Do not be tempted to round the mean off to a whole number – it is perfectly acceptable for the mean to be a decimal number, even if all the data are whole numbers (integers).

Check your understanding 11.2

1 The table shows the number of people in each of 16 cars.

Number of people, n	Frequency, f
1	7
2	5
3	3
4	1

Work out the mean number of people in a car.

2 The table shows the lengths of 20 fish in a small pond.

Length, x (mm)	Frequency, f
$0 < t \leq 30$	3
$30 < t \leq 60$	5
$60 < t \leq 90$	10
$90 < t \leq 120$	2

a) Work out the mean length of a fish.

b) State the modal class.

Comparing two distributions

Sometimes it is useful to compare the way in which two different data sets are distributed. You can see whether their averages are similar or not (use mean, median or mode). You can also see whether the spread of data is similar (use range).

Worked example

The boys and girls in class 3W took a mathematics test. The mean mark for the boys was 61, with a range of 17. The mean mark for the girls was 67, with a range of 32.

Make two comparisons between these data.

Solution

The girls on average did better than the boys (67 is higher than 61). The boys were more consistent than the girls (17 is smaller than 32).

A good visual way to compare two discrete distributions is with a **back-to-back stem-and-leaf** diagram.

Worked example

Chidera measured the heights of the ten children in her class, in centimetres:

138, 141, 147, 152, 152, 153, 157, 164, 167, 171

Richard measured the heights of the 12 children in his class, in centimetres:

144, 148, 153, 156, 159, 161, 165, 168, 171, 172, 175, 178

Illustrate these data with a back-to-back stem-and-leaf diagram.

Solution

Using stems of 130, 140, 150, 160 and 170 we have:

Chidera's class		Richard's class
8	13	
7 1	14	4 8
7 3 2 2	15	3 6 9
7 4	16	1 5 8
1	17	1 2 5 8

Key: 8 | 13 | = 138 Key: | 14 | 4 = 144

So we can see that Richard's class is taller, on average, as the leaves are generally further down the diagram than Chidera's (and therefore tend to have higher stems).

Tips for success

- Notice how the key works differently on the left-hand side of the central stems compared to the right – so it is important to include a double key as in the above example.

Check your understanding 11.3

1 The back-to-back stem-and-leaf diagram shows the ages of the members of a darts club and the members of a squash club:

Darts club		Squash club
3	1	
5 5 4	2	2
6 5 4 1	3	5 6 7 7
9 5 2	4	2 4 6 7 8
5 5 4 0	5	1 4 4 8
8 3 1	6	2
1	7	

Key: 3 | 1 | = 13 Key: | 6 | 2 = 62

Make *two* comparisons about the ages of the members of the two clubs.

Spotlight on the test

1 Here are the ages of ten children in a dance group:

10, 11, 13, 14, 15, 16, 16, 16, 17, 17

a) Work out the mean age of the children.

b) Find the median age.

c) Find the mode of the ages.

d) Work out the range of the ages.

e) A new member joins the group. Her age is 13. Without doing any new calculations, explain whether the mean age will increase or decrease. [8]

2 Here are some facts about five numbers:

- the mode is 3
- the mean is 6.2
- the median is 7
- the range is 7.

Find the values of the five numbers. [2]

3 The back-to-back stem-and-leaf diagram shows the scores obtained by two classes in their end of term mathematics test:

Mr Smith's class		Mr Hindocha's class
7 2	1	
9 8 4	2	1 8
7 3 1	3	5 6 6 9
0	4	2 5 6
	5	

Key: 2 | 1 | = 12 Key: | 2 | 1 = 21

a) Find the median score for Mr Smith's class.

b) Find the median score for Mr Hindocha's class.

c) State which class you think has done better in the test. Give a brief reason for your choice. [4]

Processing and presenting data

Pie charts

Student's book references

- Book 1 Chapter 1
- Book 2 Chapters 12 and 19
- Book 3 Chapters 12 and 19

You should be familiar with basic statistical graphs like **pie charts** and bar charts. These are used for displaying and comparing categorical data.

Worked example

The children in a tennis club are asked to vote for their favourite player of all time. Here are their replies:

Player	Number of votes
Andy Murray	12
Roger Federer	7
John McEnroe	2
Other	3

Illustrate this data in a pie chart and a bar chart.

Solution

The total number of children is $12 + 7 + 2 + 3 = 24$.
The scale is therefore one pupil $= 360 \div 24 = 15°$.

Player	Angle
Andy Murray	$12 \times 15 = 180°$
Roger Federer	$7 \times 15 = 105°$
John McEnroe	$2 \times 15 = 30°$
Other	$3 \times 15 = 45°$

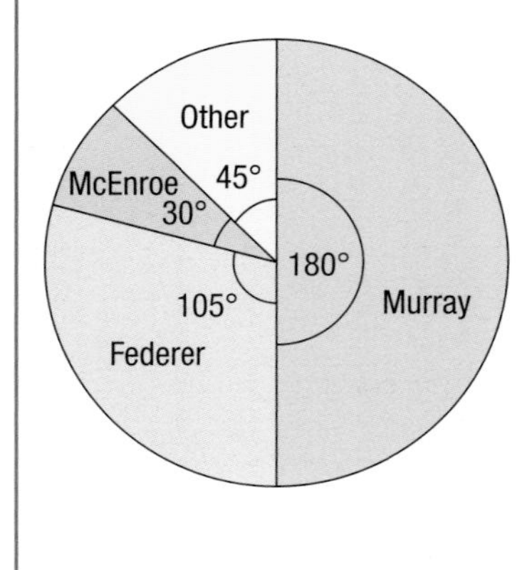

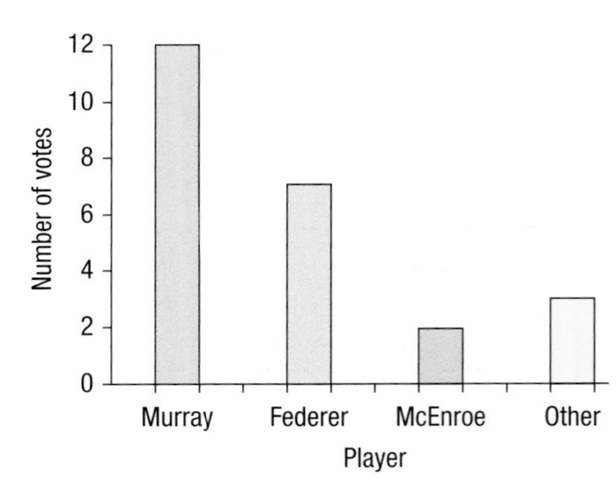

Check your understanding 12.1

1 In a class of 30 children there are five with black hair, seven with blond hair and 18 with brown hair. Illustrate this data with a pie chart.

2 Of the 20 people on a bus, eight are going to the station and seven to the cinema. The rest are going to the shopping mall. Illustrate this data with a pie chart and a bar chart.

Scatter diagrams

Scatter diagrams are used to look for **correlation** between two quantities, such as people's heights and weights. Correlation may be **positive** or **negative**, or might not even exist at all.

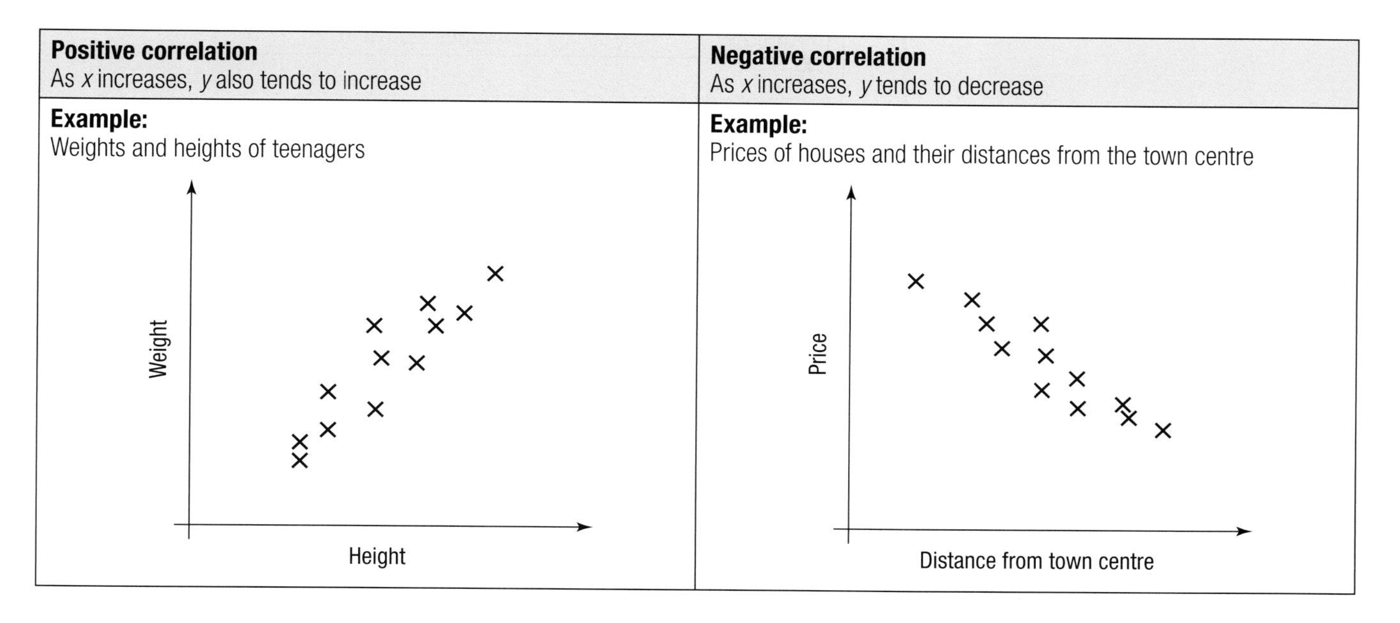

Check your understanding 12.2

1 Match the diagrams to the descriptions.

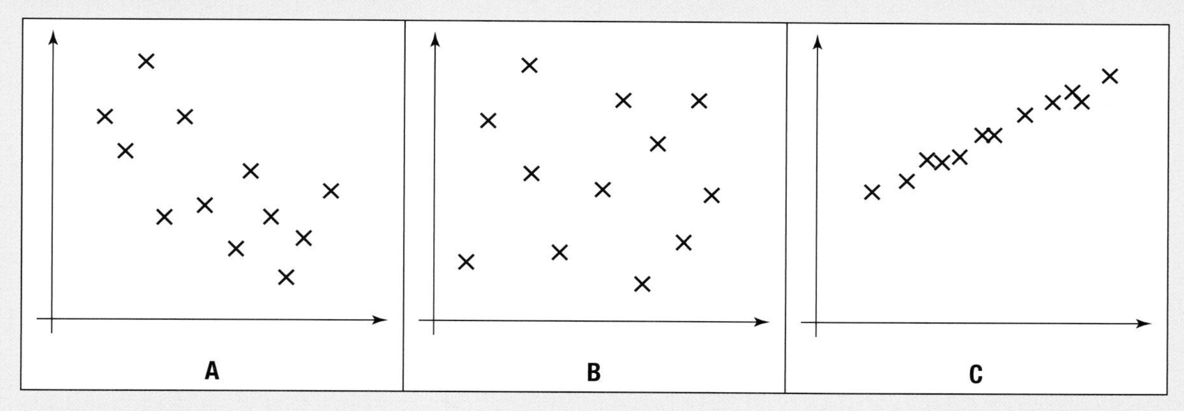

1	2	3
No correlation	**Positive correlation**	**Negative correlation**

2 In question 1, state which diagram shows *strong* correlation.

Spotlight on the test

1 Sixty vehicles are queuing for a ferry. Fifteen of them are coaches, 20 are cars and the rest are lorries. Illustrate this data with a pie chart. [3]

2 The table shows the number of certificates awarded after a recent maths challenge competition:

Certificate	Number awarded
Gold	10
Silver	15
Bronze	15
No award	50

Illustrate this data with a pie chart. [3]

3 The table shows the scores obtained by 12 students who took tests in Maths and English.

Pupil	A	B	C	D	E	F	G	H	I	J	K	L
Maths	18	15	20	16	16	15	10	8	17	18	16	13
English	15	10	19	14	10	13	6	3	13	16	12	13

a) Plot these scores on a copy of the graph below to make a scatter diagram. (The first four have been done for you.)

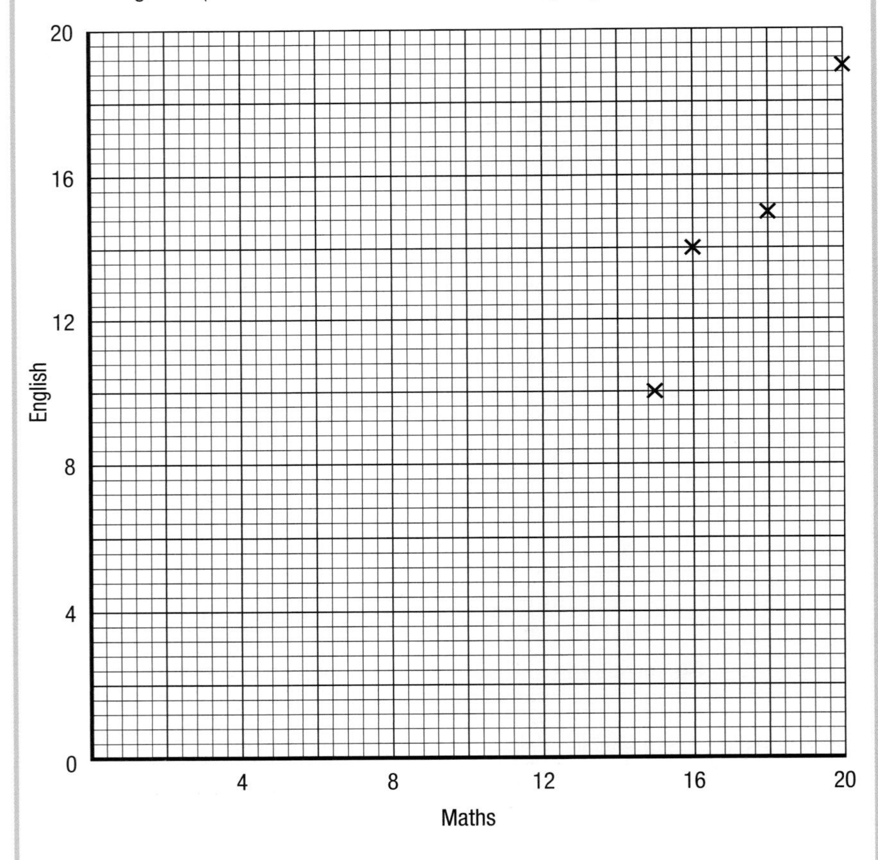

b) Describe the type of correlation between the scores.

c) Chloe scored 18 in English but was absent from the Maths test. Use the diagram to estimate what she might have scored if she had taken the Maths test. [4]

Fractions and percentages

Arithmetic with fractions

Student's book references

- Book 1 Chapter 1
- Book 2 Chapters 12 and 19
- Book 3 Chapters 12 and 19

To **add** or **subtract** fractions, write them with the same denominator.

Worked example

Work out $\frac{7}{10} - \frac{1}{4}$

Solution

$$\frac{7}{10} - \frac{1}{4} = \frac{14}{20} - \frac{5}{20}$$
$$= \frac{14 - 5}{20}$$
$$= \frac{9}{20}$$

To **multiply** two fractions, just multiply the denominators and multiply the numerators.

Worked example

Work out $\frac{3}{4} \times \frac{2}{5}$

Solution

$$\frac{3}{4} \times \frac{2}{5} = \frac{3}{2} \times \frac{1}{5}$$
$$= \frac{3 \times 1}{2 \times 5}$$
$$= \frac{3}{10}$$

Tips for success

- Some questions provide an opportunity to *cancel* a number on the top with a number on the bottom – you should always do this, as it makes the arithmetic much simpler, as in these two examples.

Worked example

Work out $\frac{3}{8} \times \frac{10}{21}$

Solution

$$\frac{3}{8} \times \frac{10}{21} = \frac{1}{8} \times \frac{10}{7}$$
$$= \frac{1}{4} \times \frac{5}{7}$$
$$= \frac{5}{28}$$

To **divide** one fraction by another, invert the second fraction and multiply.

Worked example

Work out $\frac{3}{5} \div \frac{6}{7}$

Solution

$$\frac{3}{5} \div \frac{6}{7} = \frac{3}{5} \times \frac{7}{6}$$
$$= \frac{1}{5} \times \frac{7}{2}$$
$$= \frac{7}{10}$$

When multiplying or dividing with **mixed fractions**, you must convert to top-heavy fractions first.

Worked example

Work out $1\frac{4}{5} \times 2\frac{1}{3}$

Solution

$$1\frac{4}{5} \times 2\frac{1}{3} = \frac{9}{5} \times \frac{7}{3}$$
$$= \frac{3}{5} \times \frac{7}{1}$$
$$= \frac{21}{5}$$
$$= 4\frac{1}{5}$$

Check your understanding 13.1

Work out these, writing your answers in the lowest possible terms.

1 a) $\frac{1}{6} + \frac{1}{2}$ b) $\frac{5}{8} + \frac{1}{3}$ c) $\frac{1}{4} + \frac{3}{5}$ d) $\frac{2}{3} + \frac{1}{6}$

2 a) $\frac{1}{2} - \frac{1}{4}$ b) $\frac{7}{10} - \frac{1}{2}$ c) $\frac{2}{3} - \frac{1}{10}$ d) $\frac{17}{20} - \frac{3}{5}$

3 a) $2\frac{1}{4} + 3\frac{1}{2}$ b) $2\frac{5}{6} - 1\frac{1}{3}$ c) $3\frac{3}{4} + 4\frac{5}{6}$ d) $2\frac{2}{3} - 1\frac{1}{2}$

4 a) $\frac{2}{3} \times \frac{3}{11}$ b) $\frac{5}{7} \times \frac{14}{15}$ c) $\frac{2}{9} \div \frac{5}{6}$ d) $\frac{3}{8} \div \frac{9}{10}$

5 a) $2\frac{6}{7} \times 3\frac{4}{15}$ b) $4\frac{1}{2} \times 2\frac{1}{3}$ c) $2\frac{1}{2} \div 3\frac{1}{8}$ d) $4\frac{2}{3} \div 3\frac{1}{2}$

Percentages and percentage change

Two fractions or ratios can easily be compared by writing them as percentages. To change a fraction into a percentage, multiply it by 100(%).

Worked example

Anu scores 17 out of 20 in an English test and 24 out of 30 in a mathematics test. In which test did he get the better mark?

Solution

Converting both scores to percentages:

17 out of 20 is $\frac{17}{20} \times 100 = 85\%$.

24 out of 30 is $\frac{24}{30} \times 100 = 80\%$.

So Anu got the better mark in English.

Percentage increase or decrease is often best handled by using a **multiplying factor**.

Tips for success

- Some questions ask you to *reverse* a percentage change. Again, the multiplying factor is a good approach, as in the next example.

Worked example

A camera costs \$320 plus sales tax at 20%. Find its total cost including sales tax.

Solution

$100\% + 20\% = 120\% = 1.20$ as a decimal. This is the multiplying factor. So the total cost is $\$320 \times 1.20 = \$384$.

Worked example

Claudia buys some theatre tickets. She has to pay a booking charge of 6% on top of the cost of the tickets. The total cost is \$79.50. Find the cost before the booking charge was added.

Solution

$100\% + 6\% = 106\% = 1.06$ as a decimal. This is the multiplying factor.

So (ticket cost) $\times 1.06 = \$79.50$

(ticket cost) $= \$79.50 \div 1.06 = \$75$.

Check your understanding 13.2

1 Write 32 as a percentage of 40.
2 Write 51 as a percentage of 60.
3 Find 36% of 500.
4 Increase 240 by 15%.
5 A printer costs \$85 plus 20% sales tax. Find the cost of the printer including tax.
6 My bill in a restaurant was \$41.44 including a service charge of 12%. Work out the bill before the service charge was added.

Spotlight on the test

1 Work out $\frac{1}{4} + \frac{2}{5}$. Show all of your working. [2]

2 Insert a fraction to make this calculation correct:
$3\frac{1}{4} \times \ldots = 4\frac{7}{8}$ [3]

3 A group of 60 friends were invited to a reunion, but only 48 turned up. What percentage of the group turned up? [2]

4 Luke took a mathematics test and scored 85%. His teacher says that Luke actually got 51 marks. How many marks was the test out of? [2]

5 In a sale, a shop reduces all prices by 15%. Coralie buys a computer that would normally cost $600. Find the cost of this computer in the sale. [2]

6 On New Year's Day a taxi firm increases all its fares by 65% compared to other, normal days.

a) Write down the multiplying factor for a 65% increase.

b) Nia takes a taxi ride that would normally cost $6.00. Work out how much she pays on New Year's Day.

c) Nathan takes a taxi ride on New Year's Day and pays $13.20. Work out how much he would pay on a normal day. [4]

14 Sequences, functions and graphs

Rules for number sequences

Student's book references

- Book 1 Chapter 1
- Book 2 Chapters 12 and 19
- Book 3 Chapters 12 and 19

Number sequences may be described using a **position-to-term** rule or a **term-to-term rule**.

Worked example

The nth term of a number sequence is $5n + 1$. Find the first four terms of the sequence, using this position-to-term rule.

Solution

$n = 1$ gives $5 \times 1 + 1 = 6$.

$n = 2$ gives $5 \times 2 + 1 = 11$.

$n = 3$ gives $5 \times 3 + 1 = 16$.

$n = 4$ gives $5 \times 4 + 1 = 21$.

So the first four terms are 6, 11, 16, 26.

Worked example

The first term of a number sequence is 4. Each later term is made by doubling the previous term and subtracting 3. Find the first four terms of the sequence, using this term-to-term rule.

Solution

The first term is 4.

The second term is $4 \times 2 - 3 = 5$.

The third term is $5 \times 2 - 3 = 7$.

The fourth term is $7 \times 2 - 3 = 11$.

So the first four terms are 4, 5, 7, 11.

Sequences which go up (or down) in equal steps are called **arithmetic sequences**. You can find the position-to-term rule for the nth term of such a sequence by basing it on the **position-to-term** rule or a **term-to-term rule**.

Worked example

The first five terms of a number sequence are 5, 8, 11, 14, 17. Find a rule for the nth term of this sequence.

Solution

The terms go up in steps of 3.

The rule $3n$ would generate 3, 6, 9, 12, 15, each of which is too small by 2. This must be added on to the $3n$ part of the rule.

Thus the rule for the nth term of the sequence is $3n + 2$.

Check your understanding 14.1

1 Find the next two terms in each sequence. For those that are arithmetic sequences, give a rule for the nth term.

a) 1, 3, 5, 7, 9, … .

b) 1, 2, 4, 7, 11, … .

c) 21, 25, 29, 33, … .

d) 11, 10, 9, 8, … .

2 The first term of a number sequence is 2. To make each subsequent term you multiply the previous term by 3 and add 1. Find the first four terms of the sequence.

3 The nth term of a sequence is $n^2 + 4$. Write down the first three terms.

Linear functions

If you graph an arithmetic sequence, the points will lie on a straight line: the graph is **linear**. The equation of such a straight line may be written as $y = \mathbf{m}x + \mathbf{c}$, where ***m* is the gradient** and **c is the intercept** on the y axis.

Worked examples

The diagram shows a graph of a linear function and points with $x = 1, 2, 3, 4$.

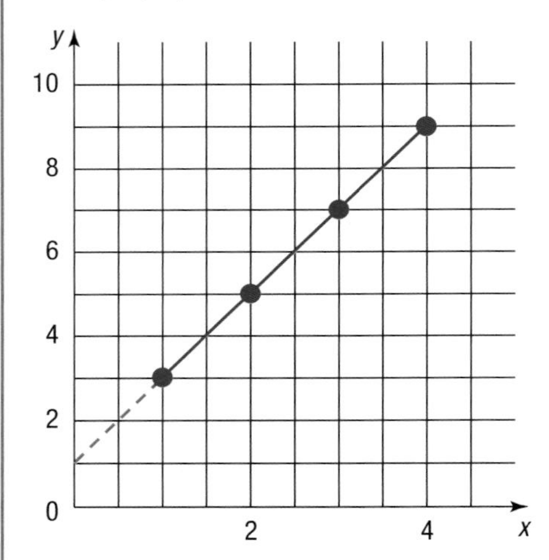

a) Find the equation of the straight line passing through these points.

b) The point $(40, k)$ lies on this line. Find k.

Solution

a) The gradient of the line is $(9 - 3) \div (4 - 1) = 6 \div 3 = 2$. This is m. The y intercept is 1. This is c. So the equation of the straight line is $y = 2x + 1$.

b) When $x = 40$, $k = 2 \times 40 + 1 = 81$.

Check your understanding 14.2

1 Find the equations of these three straight lines.

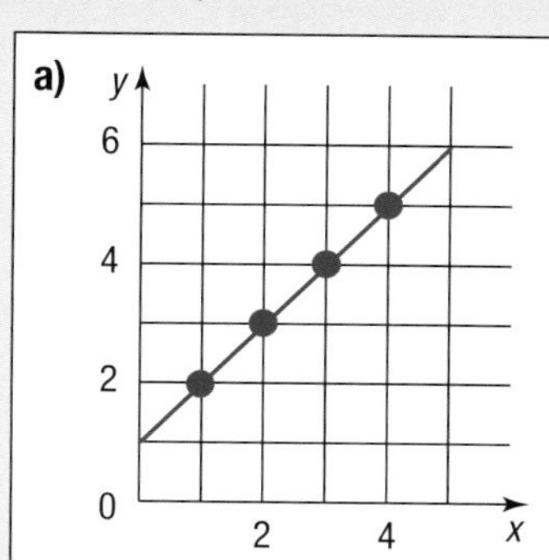

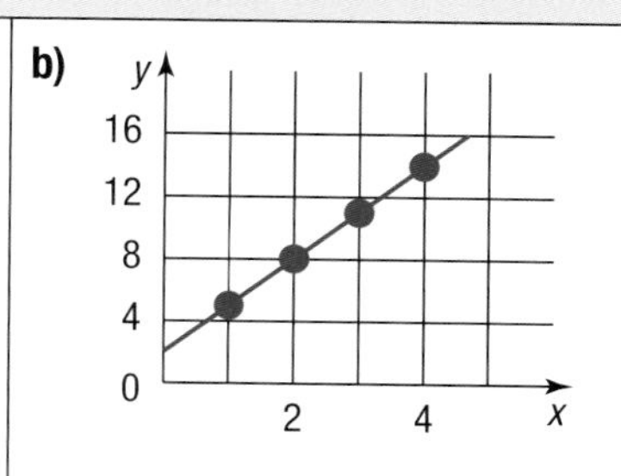

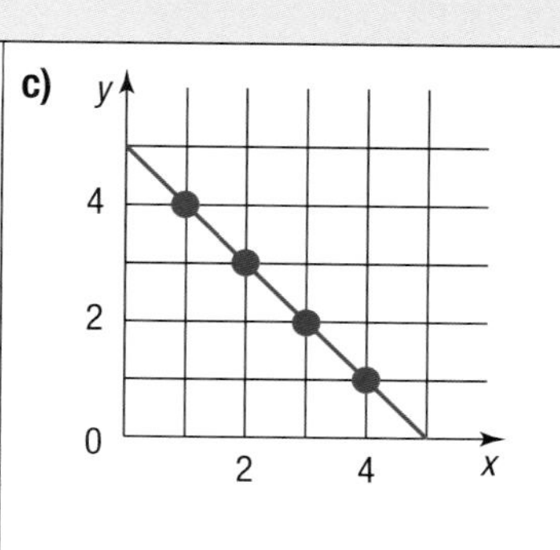

Spotlight on the test

1 Match the expression for the nth term to the correct sequence.

A	$2n + 1$	**P**	9, 8, 7, 6, …
B	$10 - n$	**Q**	1, 5, 9, 13, …
C	$n + 2$	**R**	3, 5, 7, 9, …
D	$4n - 3$	**S**	3, 4, 5, 6, …

[4]

2 The first four terms of an arithmetic sequence are 7, 10, 13, 16.

a) Find an expression for the nth term of this sequence.

b) Find the 50th term of the sequence. [2]

3 Match the linear equation to the correct graph.

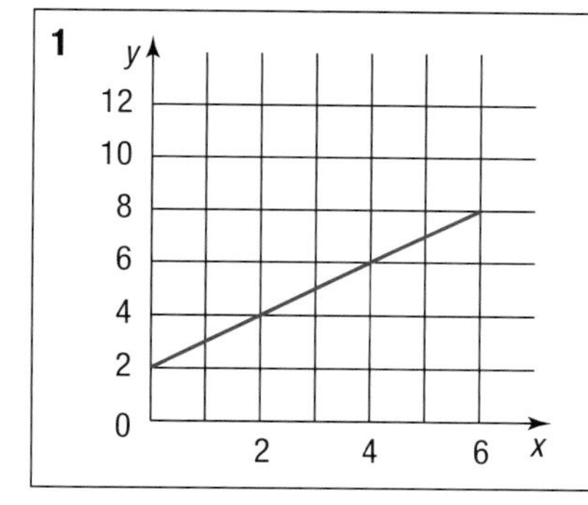

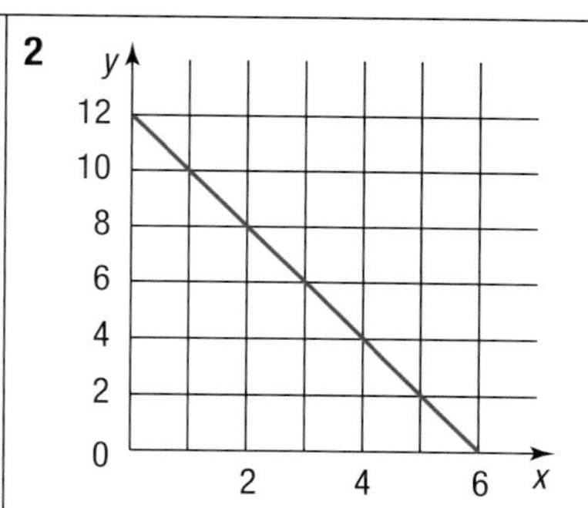

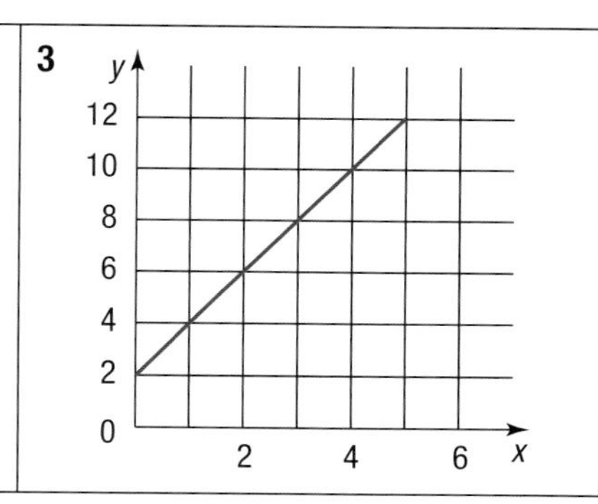

A $y = 12 - 2x$

B $y = 2x + 2$

C $y = x + 2$ [3]

4 The line **L** is parallel to $y = 3x + 2$, and passes through the point (1, 7). Find the equation of the line **L**. [2]

15 Angle properties

Angles at a point and on a line

Student's book references

- Book 1 Chapter 17
- Book 2 Chapter 17

Tips for success

- When solving angle problems, always give brief reasons for your working.

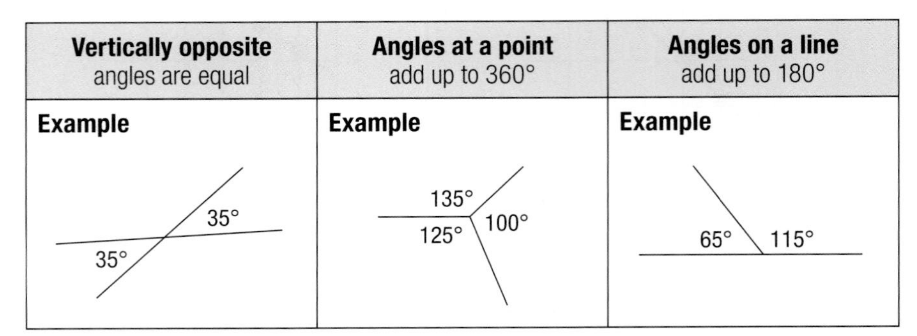

Check your understanding 15.1

1 Find each of the angles represented by letters. Give brief reasons.

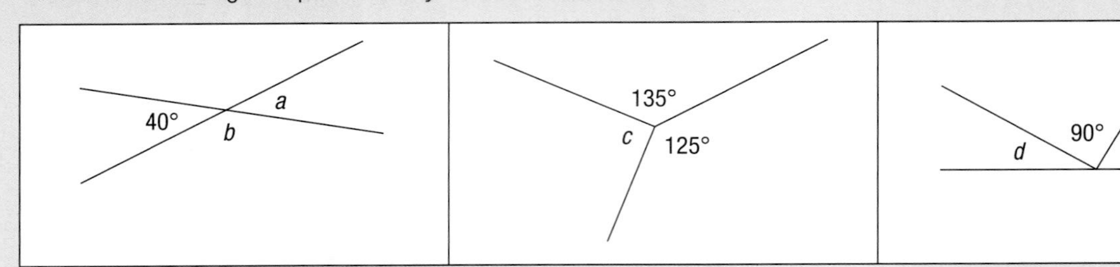

2 Find the angles indicated by letters. Explain your reasoning.

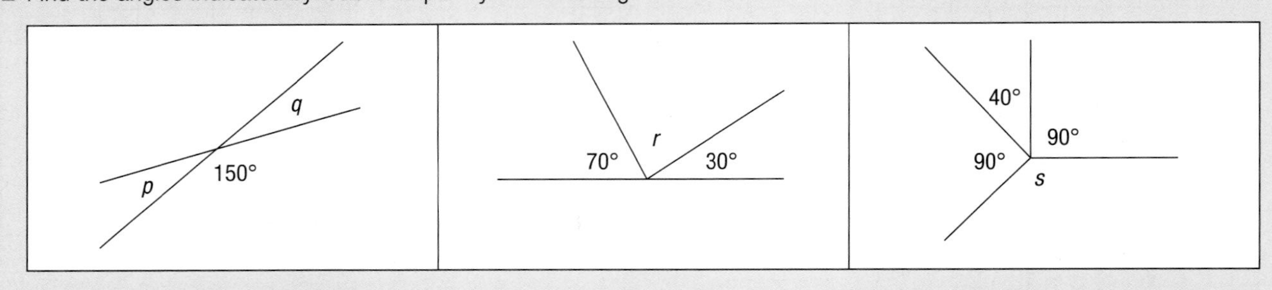

3 Find the value of each of the angles marked t in this diagram.

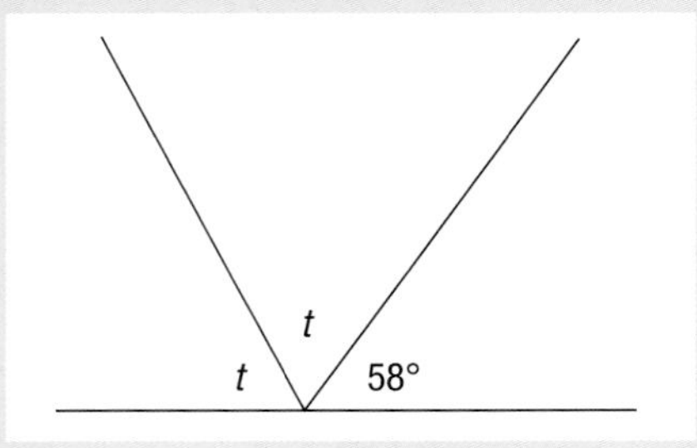

Angles and parallels

Tips for success

- Sometimes alternate angles are called Z-angles, and corresponding angles F-angles. It is best *not* to use these unofficial names in exams.

Alternate angles are equal	**Corresponding angles** are equal
Example 63° 63°	**Example** 125° 125°

Check your understanding 15.2

1 Find each of the angles represented by letters. Give brief reasons.

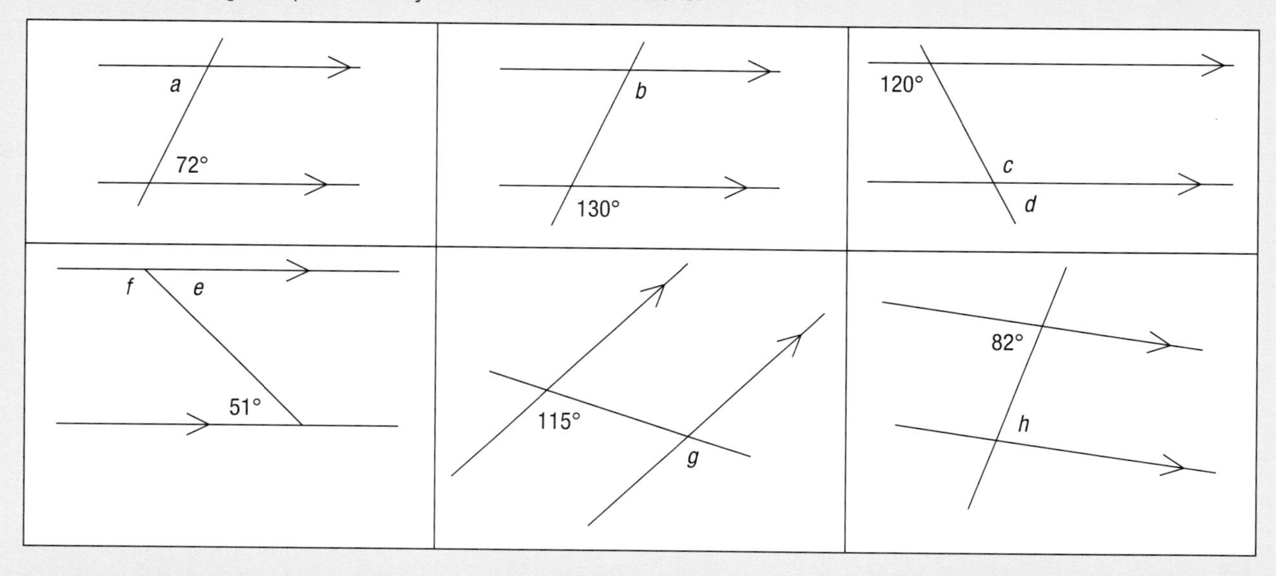

2 Copy and complete these sentences:

a) The proper name for Z-angles is ______________________ angles.

b) The proper name for F-angles is ______________________ angles.

3 Find the values of the angles marked *a* and *b* in the diagram below. Give reasons for your answers.

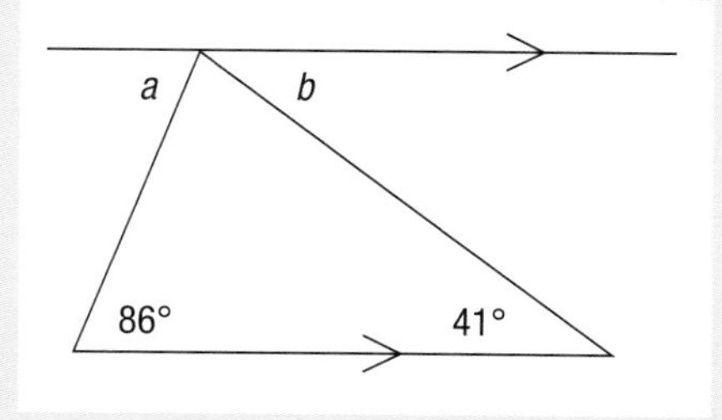

Angles in triangles and quadrilaterals

The **angles in a triangle** always add up to **180°**.
The **angles in a quadrilateral** always add up to **360°**.

Worked examples

Find the missing angles, marked m and n.

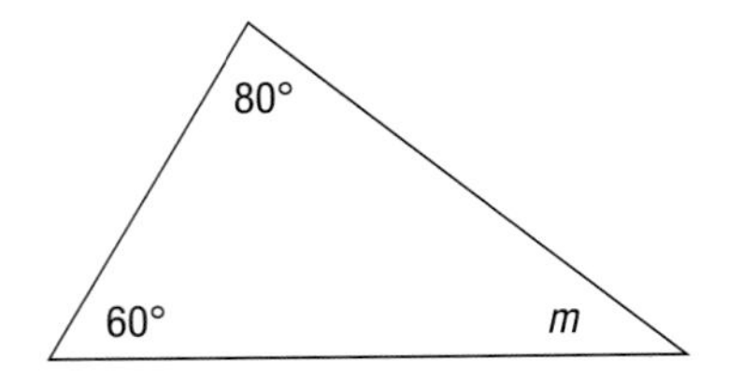

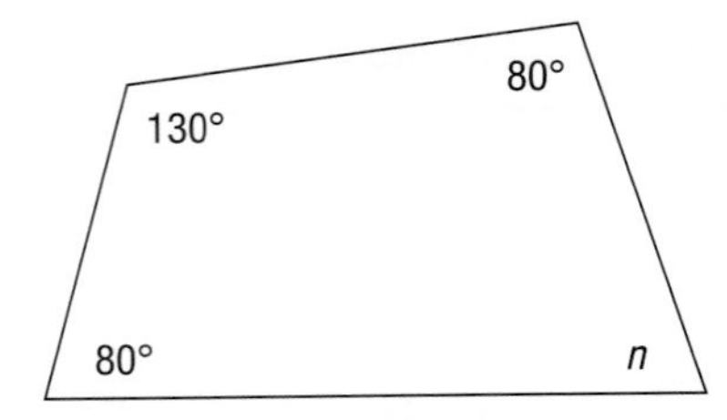

Solution

For the triangle,

$60 + 80 + m = 180°$

$140 + m = 180°$

$m = 180 - 140°$

$m = 40°$.

For the quadrilateral,

$80 + 130 + 80 + n = 360°$

$290 + n = 360°$

$n = 360 - 290°$

$n = 70°$.

Check your understanding 15.3

1 Find each of the angles represented by letters. Give brief reasons.

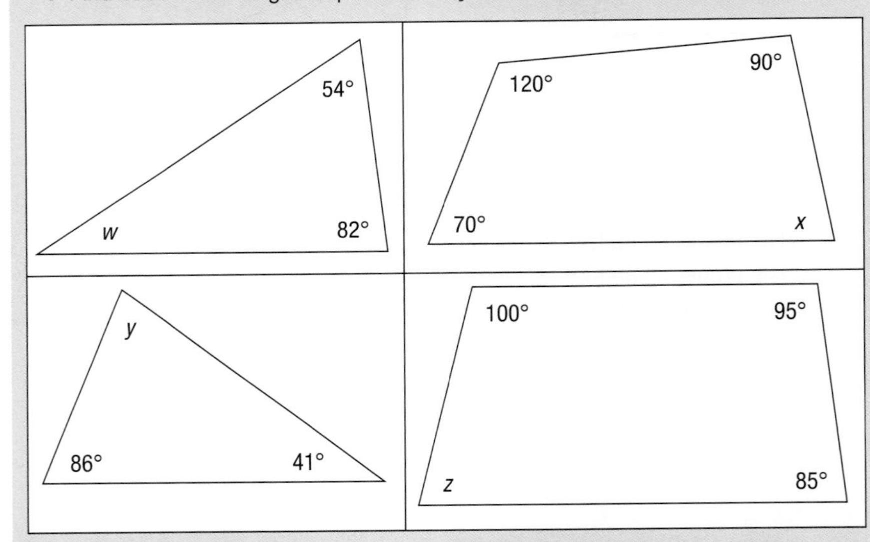

2 Two of the angles in a triangle are 53° and 87°. Work out the size of the third angle.

3 Three of the angles in a quadrilateral are 40°, 65° and 120°. Work out the size of the fourth angle.

4 Anya measures the three angles in a triangle. She obtains values of 57°, 63° and 70°. Show that Anya must have made a mistake.

Spotlight on the test

1 Find the angles marked x and y in the diagrams below. [2]

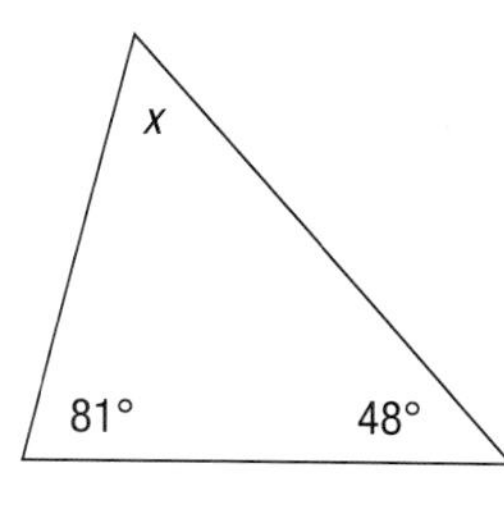

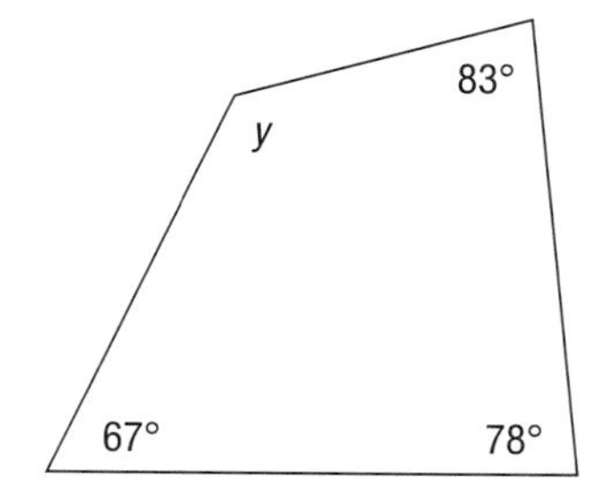

2 Complete the sentence below:

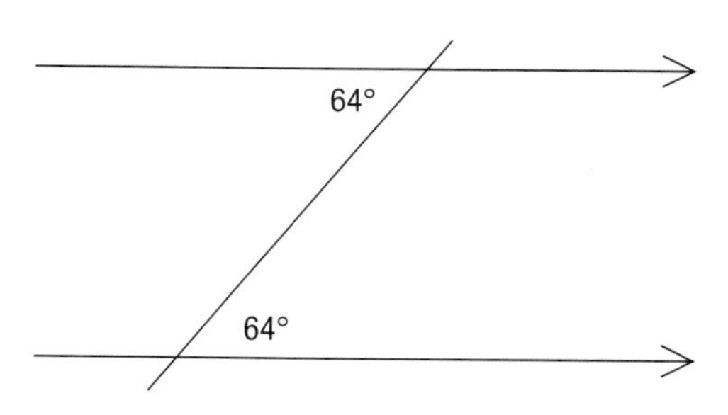

The two angles marked 64° are equal because they are _______ angles.

[1]

3 Find the angles represented by letters a to j.

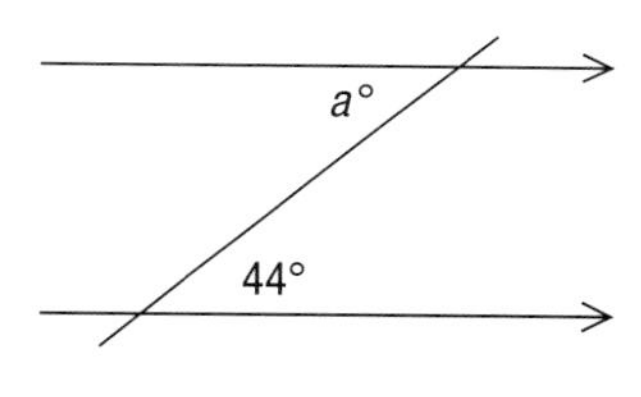

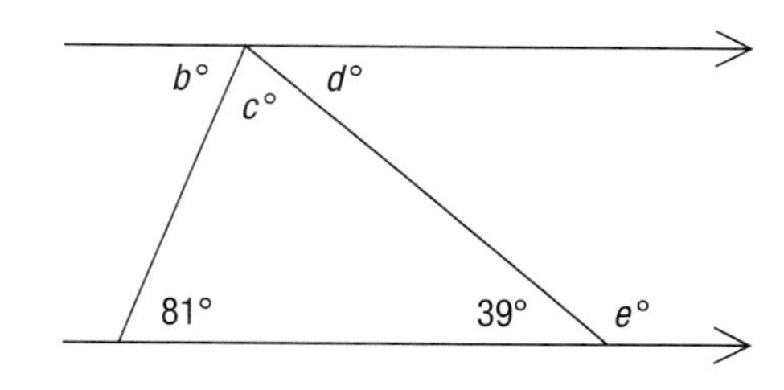

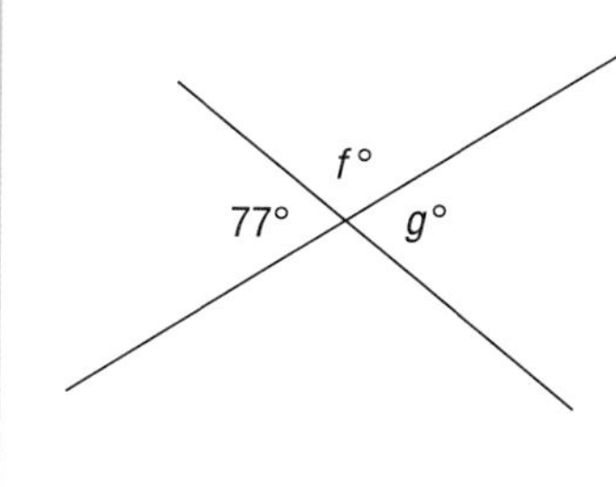

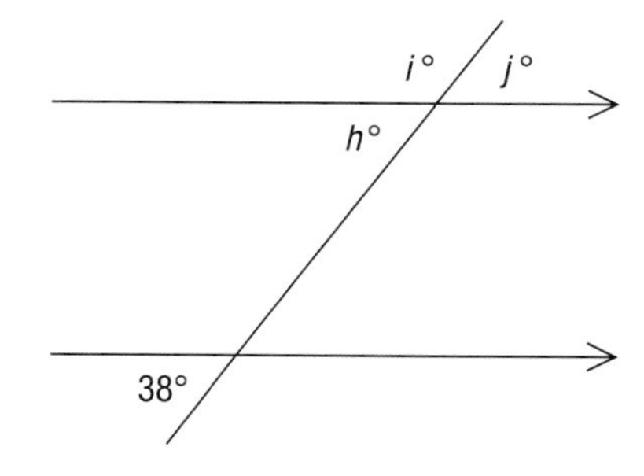

[10]

Area, perimeter and volume

Area of plane shapes: rectangle and triangle

Student's book references

- Book 1 Chapter 18
- Book 2 Chapters 18 and 25
- Book 3 Chapter 18

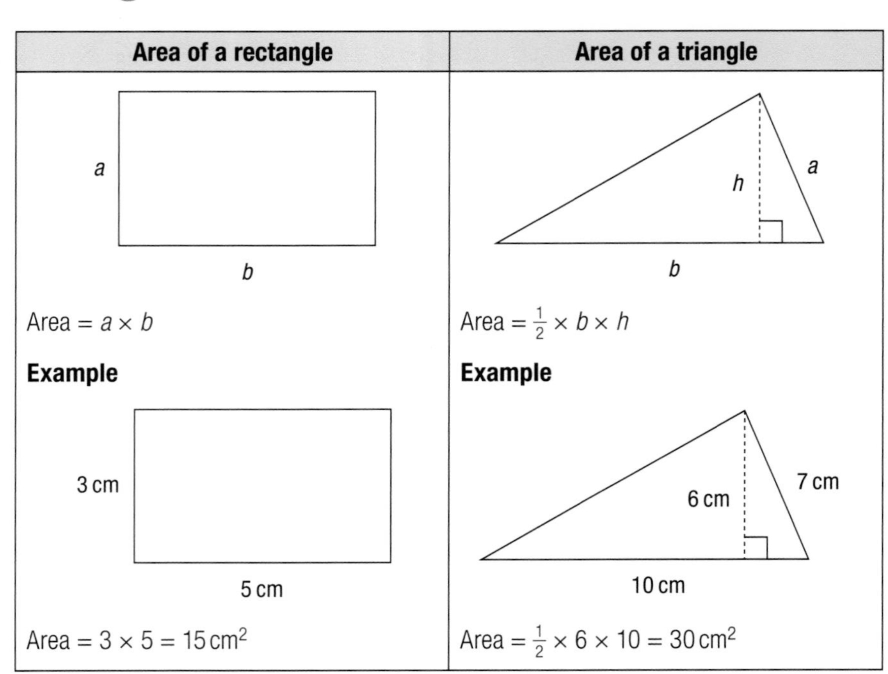

Tips for success

- Remember to include appropriate square units in your answer.

Check your understanding 16.1

Find the areas of these rectangles and triangles.

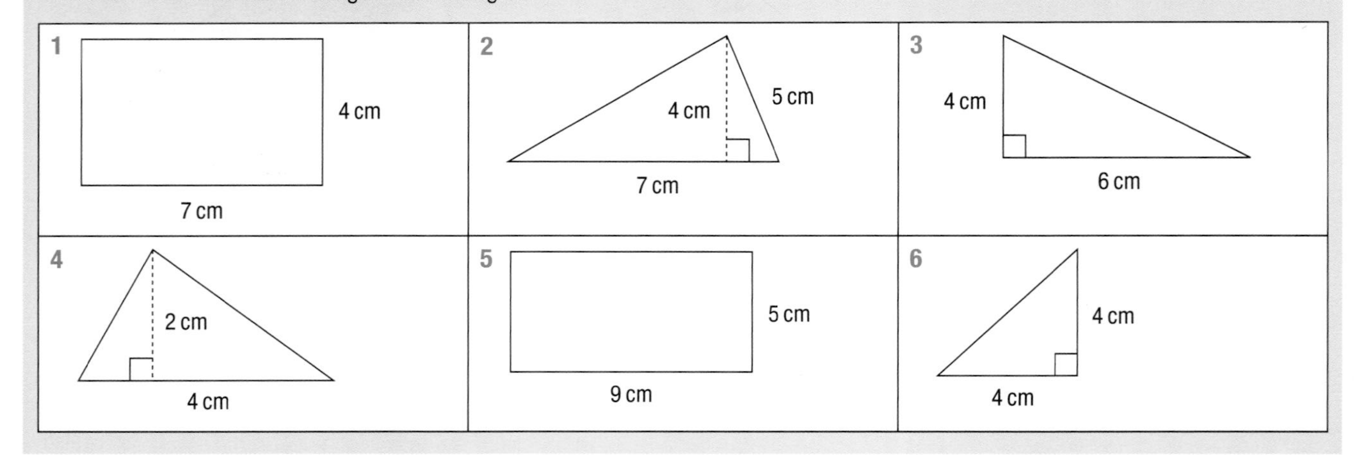

Area of plane shapes: parallelogram and trapezium

Area of a parallelogram	Area of a trapezium
a, h, b	a, h, b
Area $= b \times h$	Area $= \frac{1}{2} \times (a + b) \times h$
Example	**Example**
5 cm, 8 cm	6 cm, 4 cm, 8 cm
Area $= 5 \times 8 = 40\,\text{cm}^2$	Area $= \frac{1}{2} \times (6 + 8) \times 4 = 28\,\text{cm}^2$

Check your understanding 16.2

Find the areas of these shapes.

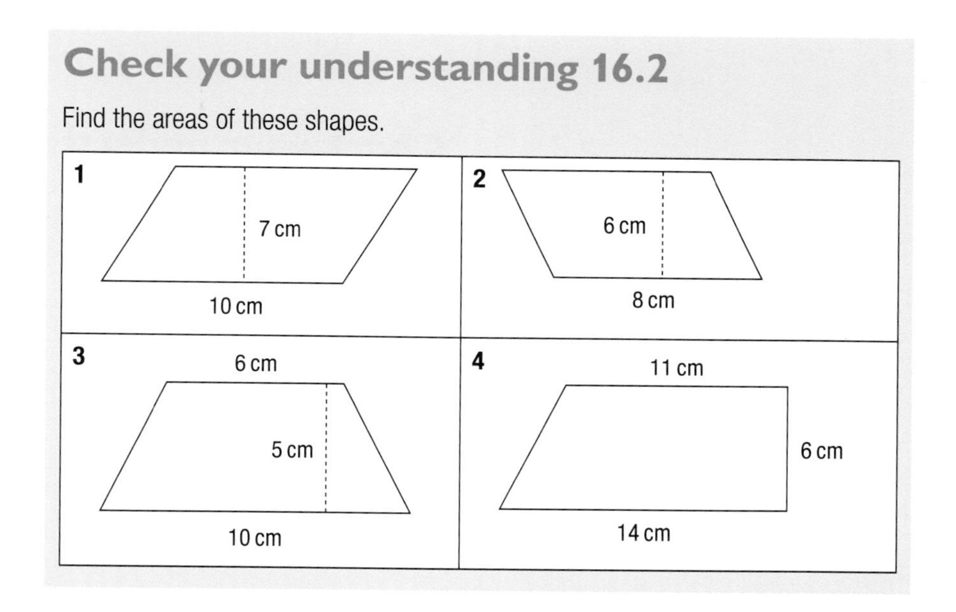

Nets

A **net** is a two-dimensional (2D) representation of a three-dimensional (3D) solid. The net can be folded to make the solid.

Worked example

What 3D solid does this net make?

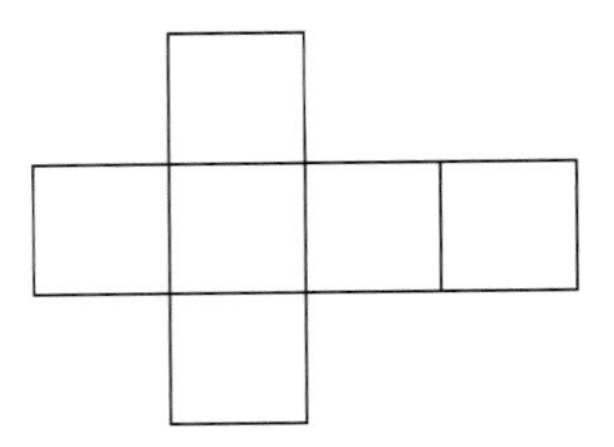

Solution

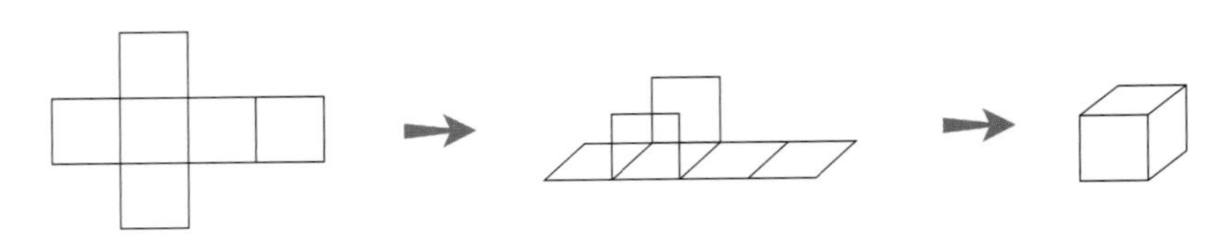

The 3D solid is a cube.

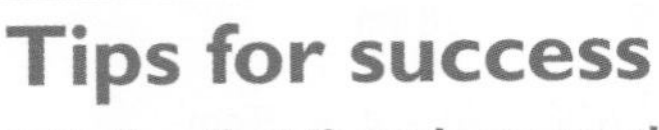

Tips for success

- Notice that there is no need to draw extra tabs, just the visible faces.

Check your understanding 16.3

1 Here are five nets. Decide which ones will make a cube.

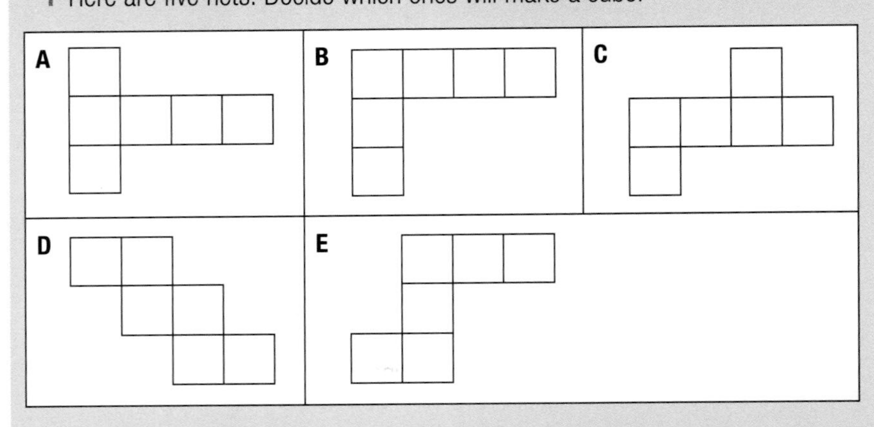

2 Here is a sketch of a 3D solid. Construct an accurate net for it.

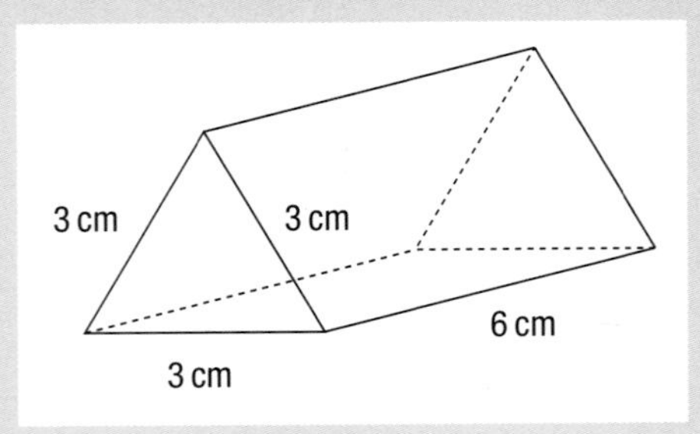

Surface area of a prism

A **prism** is a 3D solid with a constant cross-section. To find the surface area of a prism it often helps to draw a net for it first.

Worked example

Find the surface area of this prism.

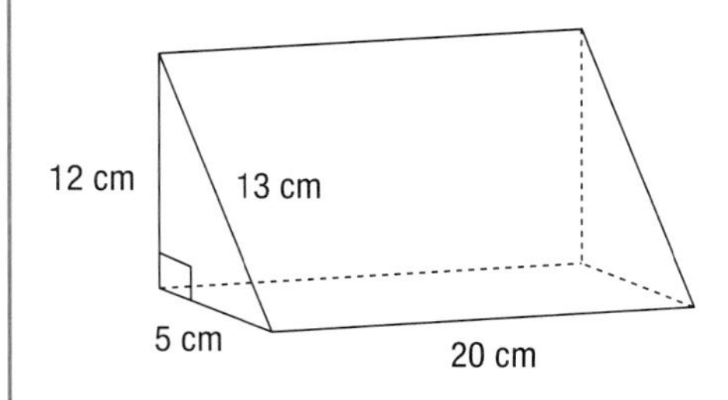

Solution

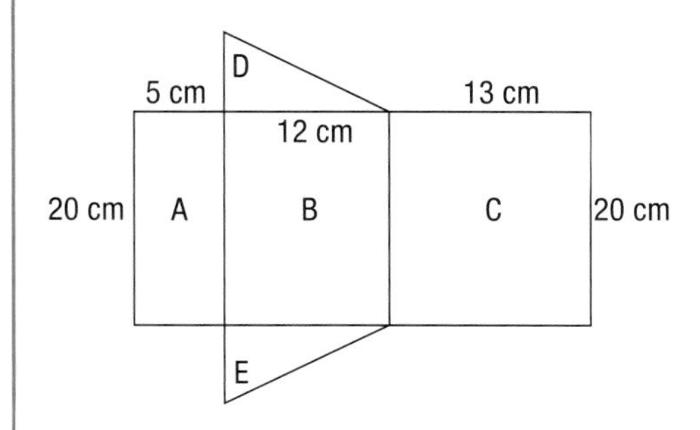

Rectangle A:	$5 \times 20 = 100\,\text{cm}^2$
Rectangle B:	$12 \times 20 = 240\,\text{cm}^2$
Rectangle C:	$13 \times 20 = 260\,\text{cm}^2$
Triangle D:	$\frac{1}{2} \times 5 \times 12 = 30\,\text{cm}^2$
Triangle E:	same as D $= 30\,\text{cm}^2$

Total surface area $= 100 + 240 + 260 + 30 + 30 = 660\,\text{cm}^2$.

Check your understanding 16.4

Find the surface area of each solid. Draw a net if it helps.

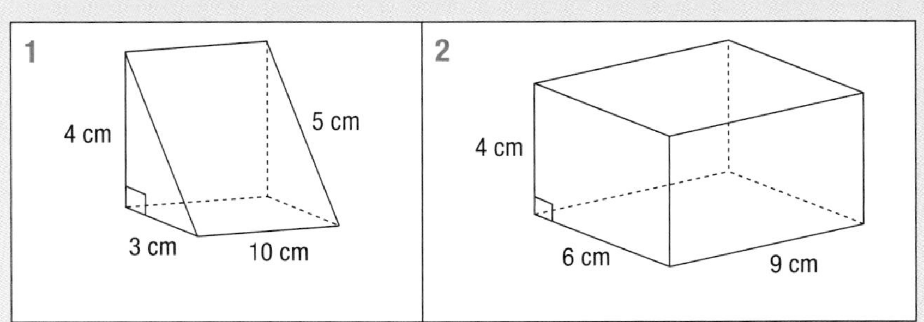

Volume of a cuboid

To find the volume of a cuboid, multiply its three dimensions together.

Worked example

Find the volume of this cuboid.

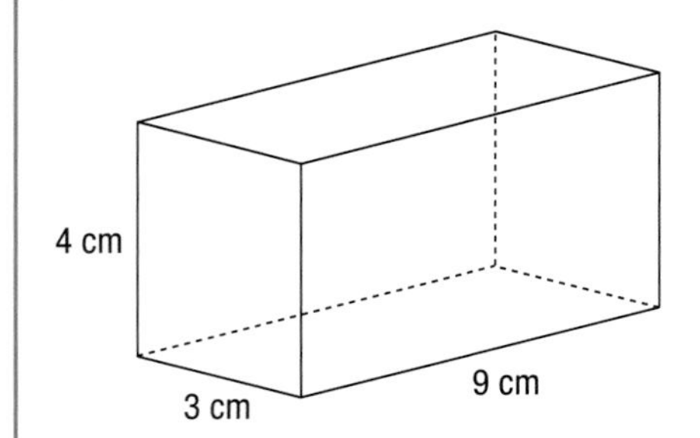

Solution

Volume = $3 \times 4 \times 9 = 108\,\text{cm}^3$.

Tips for success

- Remember to include appropriate cubic units in your answer.

Check your understanding 16.5

1 Find the volume of each cuboid.

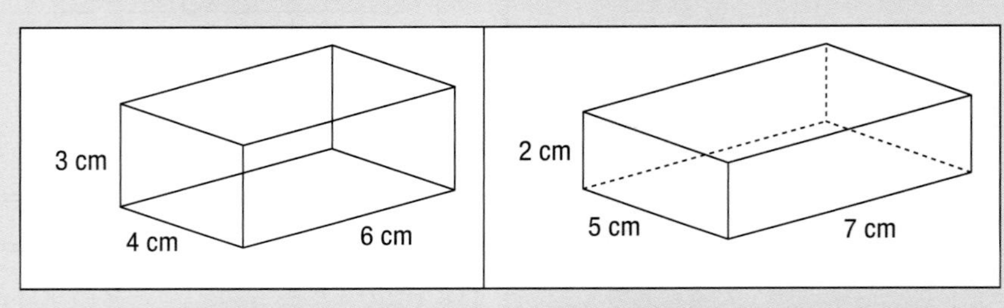

2 Each of these cuboids has a volume of $1200\,\text{cm}^3$. Find the missing lengths.

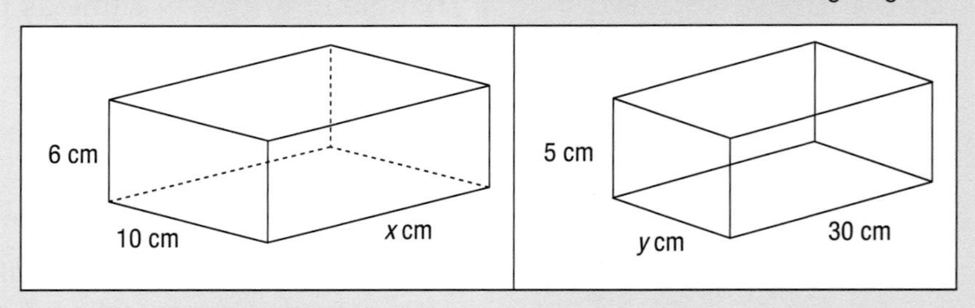

Converting metric units

Take care when converting metric units of area of volume:

$1\,\text{cm} = 10\,\text{mm}$	$1\,\text{m} = 100\,\text{cm}$
$1\,\text{cm}^2 = 10^2 = 100\,\text{mm}^2$	$1\,\text{m}^2 = 100^2 = 10\,000\,\text{cm}^2$
$1\,\text{cm}^3 = 10^3 = 1000\,\text{mm}^3$	$1\,\text{m}^3 = 100^3 = 1\,000\,000\,\text{cm}^3$

Check your understanding 16.6

1 Convert 5.5 cm^2 into mm^2.
2 The volume of a box is 0.06 m^3. Convert this into cm^3.
3 A wall has an area of 120 000 cm^2. Write this in m^2.
4 A container holds 7500 cm^3. Write this in m^3.
5 A rectangle measures 20 cm by 30 cm. Find its area in a) cm^2; b) m^2.
6 How many cubic millimetres are there in 1 cubic metre?

Spotlight on the test

1 The diagram shows a trapezium.

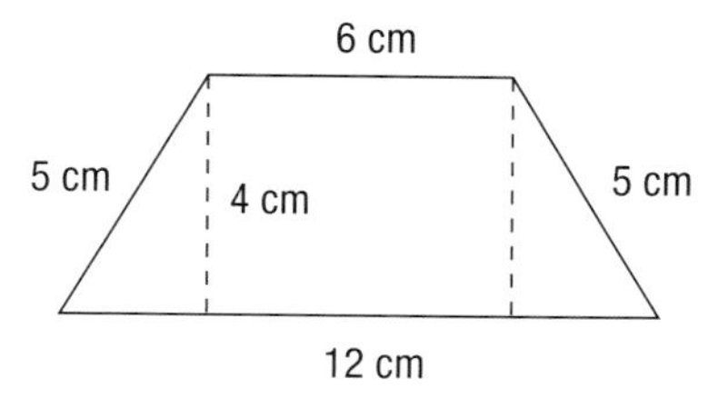

a) Calculate the perimeter of the trapezium.
b) Calculate the area of the trapezium. [4]

2 The diagram shows a 3D solid drawn on 1 cm isometric paper.

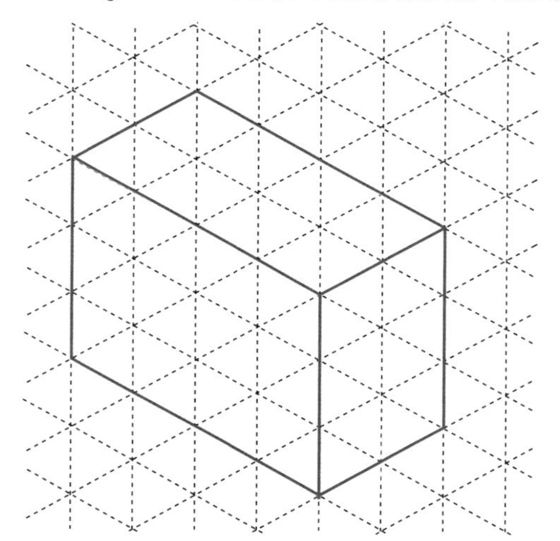

a) What name best describes this solid?
a) Construct an accurate net for the solid.
b) Calculate the total surface area of the solid.
c) Calculate the volume of the solid. [6]

3 Convert 250 cm^2 into mm^2. [1]

4 The diagram shows a triangular prism, whose cross-section is a right-angled triangle of sides 6 cm, 8 cm, 10 cm.

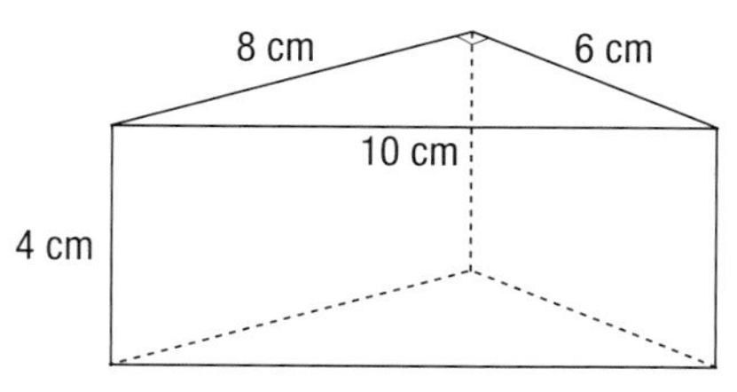

a) Sketch a net of this prism.
b) Calculate the total surface area of the prism. [3]

Ratio and proportion

Working with ratio

Student's book references

- Book 1 Chapter 22
- Book 2 Chapter 22
- Book 3 Chapter 22

A ratio indicates the relative size of two (or more) quantities. Ratios may be simplified by cancelling in a similar way to fractions.

Worked example

In a school there are 360 boys and 330 girls. Write the ratio of boys to girls in its simplest form.

Solution

The ratio is $360:330$
$= 36:33$
$= 12:11$

Worked example

Two chemicals are mixed in the ratio of $3:4$. Find the amount of each chemical, if 350 g of the mixture is made.

Solution

$3 + 4 = 7$ so divide 350 by 7 to obtain 50.

Multiplying both parts by 50, we see $3:4 = 150:200$.

So the amounts are 150 g and 200 g.

Check your understanding 17.1

1 Write each ratio in its simplest form.
a) $20:35$ b) $18:24$ c) $350:550$ d) $75:90$

2 A sheet of paper is 160 cm long. It is cut into two pieces, whose lengths are in the ratio $3:5$. Find the length of each piece.

3 In Shelley's Wood there are 200 trees. They are birch and beech trees, in the ratio $2:3$. Find the number of trees of each type.

The unitary method

Worked example

A secretary can type 200 words in 5 minutes. How long would it take him to type 840 words?

Solution

200 words take 5 minutes.

1 word takes $5 \div 200$.

840 words take $5 \div 200 \times 840 = 21$ minutes.

Tips for success

- It can be helpful to set out ratio problems using the unitary method.
- Some unitary method problems require division and multiplication to be done in the opposite order, as in this next example.

Worked example

A bag of food will last six guinea pigs for 8 days. How long will it last if there are only two guinea pigs instead?

Solution

Six guinea pigs can last for 8 days.

One guinea pig can last for 8×6 days.

Two guinea pigs can last for $8 \times 6 \div 2 = 24$ days.

Check your understanding 17.2

1. It takes Fiona 28 minutes to put labels inside 84 library books. How long would it take her to put labels inside 120 books?
2. When Stephen goes hiking he can travel 5 kilometres in 75 minutes. How long would it take him to travel 8 kilometres?
3. Marcus the postman takes 3 hours to deliver the post to 140 houses. How long would it take to deliver the post to 175 houses?
4. A supply of fuel at the bus depot can supply 12 buses for 5 days. For how long could it supply ten buses?
5. It takes four builders 6 hours to put up some scaffolding. How long would it take three builders to do the same job?

Proportion

If two quantities are always in a fixed ratio to each other then they are said to be in (direct) proportion. Simple proportion problems can be solved using ratios.

Worked example

The cost of a computer is proportional to the speed of the processor. A computer with a 2.0 GHz processor costs \$200. Find the cost of a computer with a 2.4 GHz processor.

Solution

The ratio is $2.0 : 2.4$

$= 20 : 24$

$= 200 : 240$

So the computer would cost \$240.

Check your understanding 17.3

1 A minibus is travelling along a stretch of motorway. The distance it travels is proportional to the time for which it travels. In 20 minutes it travels 30 kilometres. How far will it travel in 30 minutes?

2 The mass of a piece of steel pipe is directly proportional to its length. A piece 80 cm long has a mass of 5 kg. Find the mass of a piece of pipe 200 cm long.

3 The cost of placing an advert in a newspaper is directly proportional to the number of lines it takes up. An advert of five lines will cost $80. Find the cost of an advert of 12 lines.

4 The cost of a carpet is directly proportional to the area. An area of 10 m^2 would cost $250. Find the cost of a carpet with an area of 35 m^2.

Spotlight on the test

1 Write the ratio 48 : 60 in its simplest terms. [1]

2 A prize of $300 is shared into two parts, in the ratio 4 : 11. Find the value of the smaller share. [2]

3 A fruit drink is made by mixing squash and water in the ratio 2 : 9.

a) How many litres of water are mixed with 8 litres of squash?

b) How many litres of squash are needed to make 33 litres of the drink? [2]

4 Here is the exchange rate between the British pound £ and the euro €:

£1 = €1.20

a) Convert £250 into euros.

b) Convert €450 into pounds. [2]

5 The time Nik spends awake to the time he spends asleep is in the ratio 5 : 3. Work out the number of hours he spends asleep in any 24-hour period. [2]

6 Fifty identical toy bricks weigh a total of 400 grams. Find the total weight of 90 similar bricks. [2]

7 A gardener has enough petrol to run two lawnmowers for 80 minutes. For how long could he run five lawnmowers with the same amount of fuel? [2]

8 The cost of buying a necklace with your name on it is directly proportional to the number of letters on it. Megan buys a necklace and pays $4.80. How much would her friend Lorraine have to pay? [2]

18 Formulae, functions and graphs

Substituting into formulae

Student's book references

- Book 1 Chapter 23
- Book 2 Chapter 23
- Book 3 Chapter 23

Tips for success

- Some substitution problems require rearrangement after putting the numbers in, so that the required quantity becomes the subject. The second example shows you how this is done.

You can work out the value of a quantity by substituting into a formula. Take care to follow BIDMAS (see page 11) correctly.

Worked example

The quantities v, u, a, t are related by the formula $v = u + at$. Work out the value of v when $u = 15$, $a = 10$ and $t = 4$.

Solution

$v = u + at$
$= 15 + 10 \times 4$
$= 15 + 40$
$= 55.$

Worked example

The quantities F and C are related by the formula $F = 2C + 30$. Work out the value of C when $F = 100$.

Solution

$F = 2C + 30$
$100 = 2C + 30$
$2C + 30 = 100$
$2C = 70$
$C = 35.$

Check your understanding 18.1

1 $A = ky + 10$. Find A if $k = 7$ and $y = 2$.
2 $L = m(10 + n)$. Find L if $m = 100$ and $n = 3$.
3 $T = 5a + 2bc$. Find T when $a = 7$, $b = 2$ and $c = -3$.
4 $Q = a^2 + b^2 + c^2$. Find the value of Q when $a = 3$, $b = 4$ and $c = 5$.
5 $y = x^2 - 4x + 5$. Find y when $x = 6$.
6 $F = ma$. Find F when $m = 12$ and $a = 10$.
7 $y = 6x + 3$. Find x when $y = 27$.
8 $V = abc$. Find the value of c when $V = 240$, $a = 6$ and $b = 8$.
9 $s = ut + 5t^2$. Find u when $s = 210$ and $t = 6$.
10 $PV = RT$. Find P when $V = 100$, $R = 40$ and $T = 70$.

Trial and improvement

Some equations are too hard to solve by exact methods. You can use a method of trial and improvement to produce an approximate solution, correct to a specified level of accuracy.

Worked example

The equation $x^2 + x - 5 = 0$ has a solution between $x = 1$ and $x = 2$. Use trial and improvement to find this solution correct to one decimal place.

Solution

When $x = 1$, then $1^2 + 1 - 5 = -3$ (which is too low)
When $x = 2$, then $2^2 + 2 - 5 = +1$ (which is too high).
Now try using one decimal place:
When $x = 1.6$, then $1.6^2 + 1.6 - 5 = -0.84$ (which is too low)
When $x = 1.7$, then $1.7^2 + 1.7 - 5 = -0.41$ (which is still too low)
When $x = 1.8$, then $1.8^2 + 1.8 - 5 = +0.04$ (which is too high).
So the solution lies between 1.7 and 1.8, and clearly seems much closer to 1.8. Check this decision, by using 1.75:
When $x = 1.75$ then $1.75^2 + 1.75 - 5 = -0.1875$ (which is too low)
So the solution is indeed $x = 1.8$ (correct to one decimal place).

Tips for success

- Be sure to include the details of the trials within your written answer.

Check your understanding 18.2

1 The equation $x^2 + 3x - 7 = 0$ has a solution between $x = 1$ and $x = 2$. Find this solution correct to one decimal place.
2 The equation $x^2 - 7x - 9 = 0$ has a solution between $x = 8$ and $x = 9$. Find this solution correct to one decimal place.
3 The equation $x^2 - x - 5 = 0$ has a solution between $x = 2$ and $x = 3$. Find this solution correct to one decimal place.
4 The equation $x^2 + 5x - 5 = 0$ has a solution between $x = 0$ and $x = 1$. Find this solution correct to one decimal place.
5 The equation $x^2 - x - 10 = 0$ has a solution between $x = 3$ and $x = 4$. Find this solution correct to one decimal place.
6 The equation $x^2 - 6x + 7 = 0$ has a solution between $x = 1$ and $x = 2$. Find this solution correct to one decimal place.
7 The equation $x^3 - 8x + 3 = 0$ has a solution between $x = 2$ and $x = 3$. Find this solution correct to one decimal place.
8 The equation $x^2 - 4x + 1 = 0$ has a solution between $x = 0$ and $x = 1$. Find this solution correct to one decimal place.
9 The equation $x^2 = 10 - x$ has a solution between $x = 2$ and $x = 3$. Find this solution correct to one decimal place.
10 The equation $x^3 = 12x - 7$ has a solution between $x = 3$ and $x = 4$. Find this solution correct to one decimal place.

Gradient and intercept of a straight line

Recall that the standard equation of a straight line is $y = mx + c$, where m is the gradient and c the y intercept. Some questions require that you rearrange a given equation into this form.

Worked example

Find the gradient and intercept of the line $2x + 3y = 6$.

Solution

$2x + 3y = 6$

$3y = -2x + 6$

$y = -\frac{2}{3}x + 2$

So $m = -\frac{2}{3}$ and $c = 2$.

The gradient is $-\frac{2}{3}$ and the intercept is 2.

Check your understanding 18.3

Rearrange each of these into the form $y = mx + c$ and hence find the gradient and intercept in each case.

1 a) $y - 3x - 4 = 0$ **b)** $\frac{y}{3} - x = 0$ **c)** $y + 4x - 2 = 0$

2 a) $3y - 6x = 4$ **b)** $4y - x + 3 = 0$ **c)** $3x + 2y = 12$

3 a) $x + y = 10$ **b)** $3x - 4y + 5 = 0$ **c)** $4x + 3y - 1 = 0$

Modelling real-life problems

By forming and solving equations from given information, it is possible to solve equations about real-life situations.

Worked example

The cost $\$C$ of running a charity theatre trip for n people is given by the formula $C = 70 + 35n$.

a) Complete this table of values.

n	0	5	10	15
C		245		

b) Illustrate the data with a graph.

c) The organisers have a limit of \$500 available for the trip. Use your graph to determine the maximum number of people who could go on the trip.

Solution

a)

n	0	5	10	15
C	70	245	420	595

b)

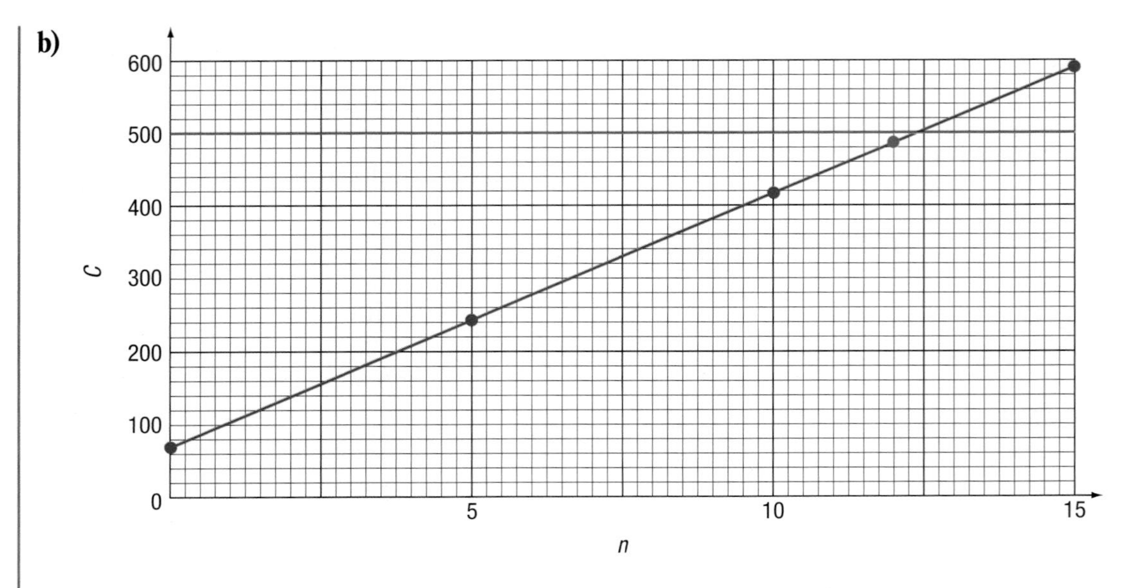

c) From the graph we can see that the point with $n = 12$ lies below the line $y = 500$, while the point for $n = 13$ lies above it.

So the maximum number of people is 12.

Check your understanding 18.4

1 A van hire firm has two price plans:

Plan A: $C = 40 + 0.4n$

Plan B: $C = 100 + 0.25n$

C represents the total hire cost, in \$, when you drive n kilometres. These two plans are illustrated on the graph below.

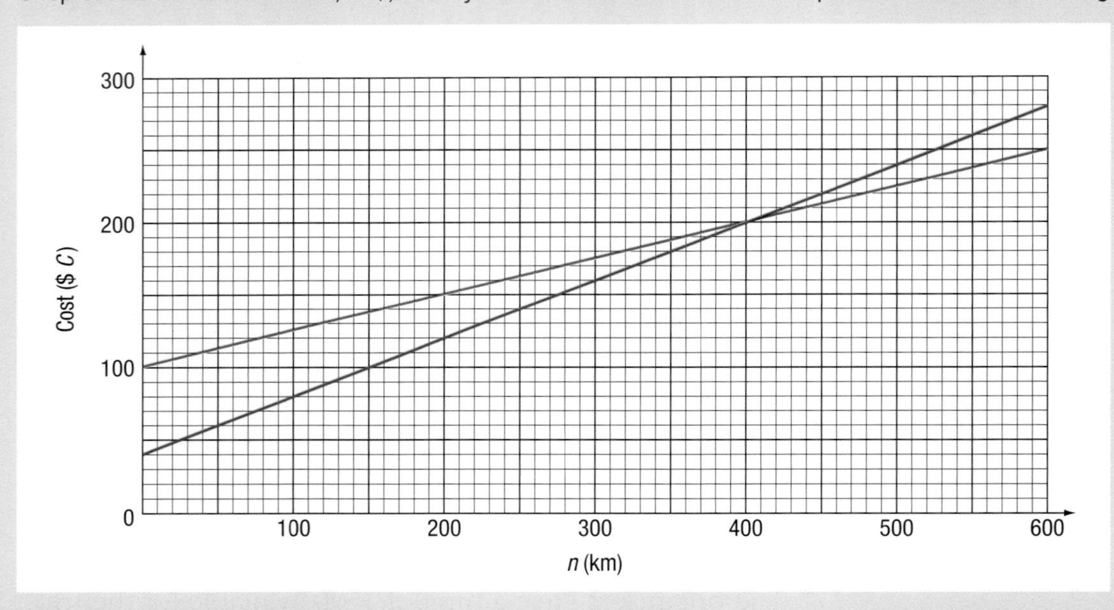

a) State which plan corresponds to the red line.

b) Martha drove 500 km. Which plan should she use?

c) Fred drove x km, and found the price was the same under both plans. Find the value of x.

Spotlight on the test

1 x and y are connected by the equation $y = 5x + 7$.

a) Find the value of y when $x = 4$.

b) Find the value of x when $y = 52$. [3]

2 E and r are connected by the equation $E = mr^2$.

a) Find the value of E when $m = 6$ and $r = 10$.

b) Find the value of r when $E = 200$ and $m = 8$. [3]

3 Show that the equation $x^3 = 64 + x$ has a solution between $x = 4$ and $x = 5$. Use trial and improvement to find its value correct to one decimal place. [3]

4 A straight line has equation $6x - 5y = 15$.

a) Write this equation in the form $y = mx + c$.

b) State the gradient of the line.

c) State the y intercept of the line. [3]

5 Here are the equations of five straight lines:

P $y = 3x + 2$

Q $x + y = 10$

R $y = 6 - x$

S $3y - x = 15$

T $2y - 4x = 7$.

a) Which line has gradient 2?

b) Which two lines are parallel?

c) Which line passes through the point (0, 5)? [3]

6 Two electric cars A and B are racing. The graph shows their distance s metres, from the start point, after t seconds.

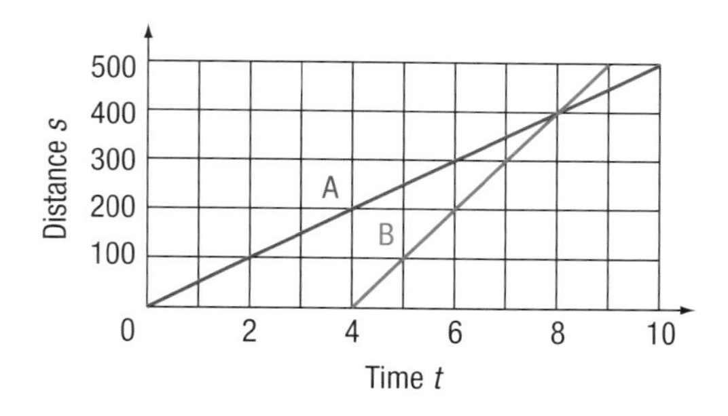

a) One of the cars had difficulty getting started. Which one?

b) One of these graphs has equation $s = 50t$. Which one?

c) At what time does one car overtake the other?

d) The racetrack is 600 metres in length. Which car is likely to be the winner? Explain your answer. [4]

19 Bearings and drawings

Maps and scales

Student's book references

- Book 1 Chapter 23
- Book 2 Chapter 23
- Book 3 Chapter 23

A **map** is simply a **scale drawing** of the main features of a landscape. The **scale** of a map may be stated either as a ratio, like 1 : 25 000, or as a key, like 1 cm = 1 km. You need to be able to convert between these two forms.

Worked example

A map has been made to a scale of 1 : 25 000. On the map two towns are 8 cm apart. How far apart, in km, are the real towns?

Solution

1 : 25 000 means 1 cm = 25 000 cm

So 1 cm = 250 m

So 8 cm = 8 × 250 = 2000 m = 2 km.

Worked example

Two places are 10 km apart. On a map they are 5 mm apart. Find the scale of the map, in the form 1 : n.

Solution

The scale is 5 mm : 10 km

that is 5 mm : 10 000 m

that is 5 mm : 10 000 000 mm

that is 5 : 10 000 000

which is 1 : 2 000 000.

Check your understanding 19.1

1 A toy soldier is made to a scale of 1 : 12. The toy is 14 cm tall. How tall would the corresponding real soldier be?

2 Two towns are 120 km apart. They are shown on a map with a scale of 1 : 5 000 000. How far apart are the towns on the map?

3 A map is made to a scale of 2 cm = 1 km. Write this scale as a ratio in the form 1 : n.

4 A model of the world's tallest horse, Big Jake, has been made on a scale of 1 : 20. The real horse is 2.1 m tall. How tall is the model horse?

5 An orienteering map is made to a scale of 1 : 5000. Two features are 5 cm apart on the map. How far apart are they in reality?

Bearings

A **bearing** is a way of measuring direction on a map. Angles are measured clockwise from north (N):

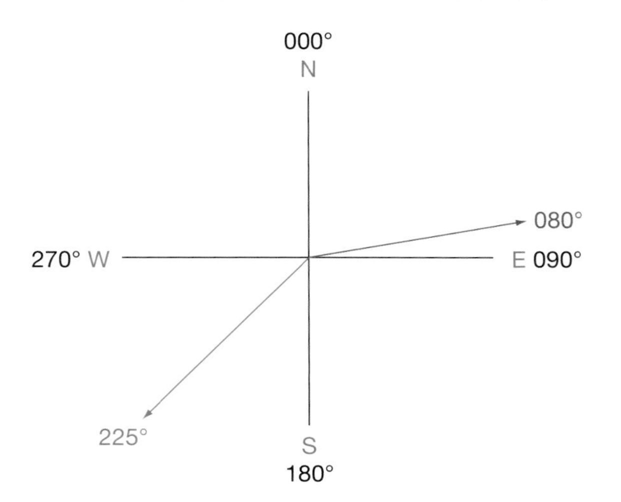

> **Tips for success**
> - Bearings should always be written using three figures, so we write 080, not just 80.

The diagram above shows bearings of 080° (blue) and 225° (red).

Check your understanding 19.2

1 A hiker plans a long walk. It consists of six stages, *A*, *B*, *C*, *D*, *E* and *F*, shown on the map. Measure the bearing of each stage, using a protractor.

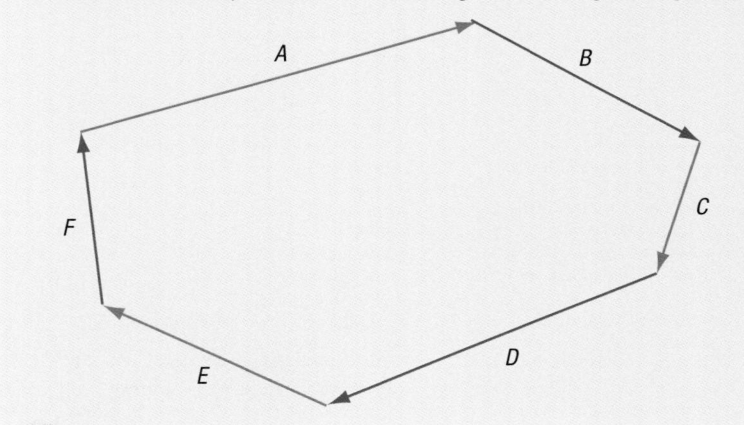

Loci

A **locus** is the path traced out when a point moves according to a given rule. Here are three of the most common ones.

P is at a fixed distance from a given point	*P* is at a fixed distance from a line	*P* is equidistant from a pair of lines
The locus is a circle	The locus is a pair of straight lines, with curving ends	The locus is a straight line

Check your understanding 19.3

1 The sketch shows a walled garden. It is in the shape of a rectangle with corners at *A*, *B*, *C*, *D*.

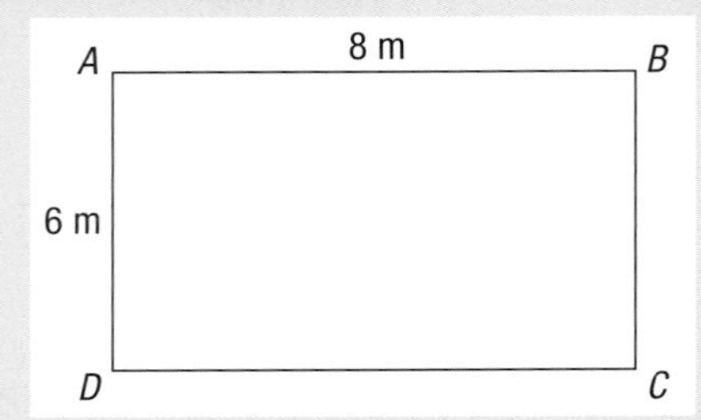

a) Make a scale drawing of the garden. Use a scale of 1 cm = 1 m.

b) All the points within 4 metres of the corner *A* are to be made into a flowerbed. Draw this flowerbed on your scale drawing.

c) A gravel path is to be laid everywhere within 2 metres of the wall *BC*. Draw the gravel path on your scale drawing.

2 Draw a square of side 5 cm.

a) Draw the locus of all the points *inside* the square that are exactly 1 cm from the perimeter of the square. Label this locus *M*.

b) Shade the region occupied by all of the points *outside* the square that are within 1 cm of the perimeter of the square. Label this region *R*.

Spotlight on the test

1 A model aircraft has a wingspan of 1405 mm. The scale of the model is 1 : 8. What is the wingspan of the full-size aircraft? [2]

2 The sketch shows a plan of a room.

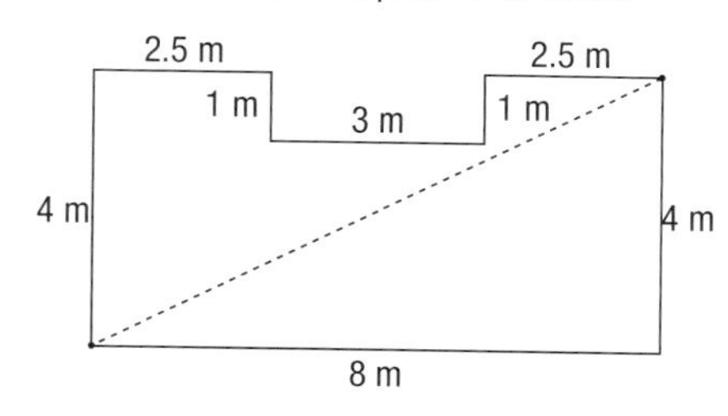

a) Make a scale drawing of the room. Use a scale of 1 cm = 1 m.

b) Measure your diagram to find the length of the longest diagonal of the room (marked with a dashed line on the sketch). [3]

3 The diagram shows a map, on a 1 cm grid, of a small island and two villages marked *A* and *B*. The scale of the map is 1 cm to 1 km.

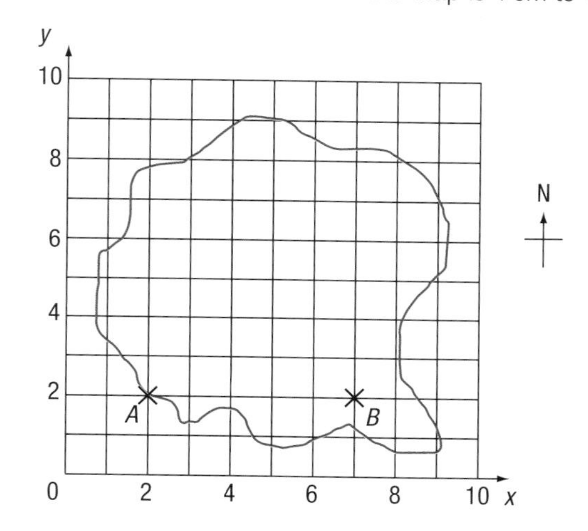

Another village *V* is on a bearing of 045° from village *A*. It is due north of village *B*.

a) Mark the position of village *V* on a copy of this map.

b) State the distance, in km, from village *B* to village *V*.

c) Measure the distance, in km, from village *A* to village *V*.

d) Write the scale of this map in the form 1 : *n*. [4]

Circles, cylinders and prisms

Area and circumference of a circle

Student's book references

- Book 1 Chapter 23
- Book 2 Chapter 23
- Book 3 Chapter 23

The area of a circle of radius r is given by $A = \pi r^2$.
The circumference of a circle of radius r is given by $C = 2\pi r$.

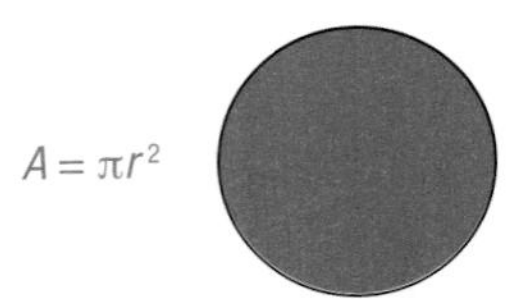

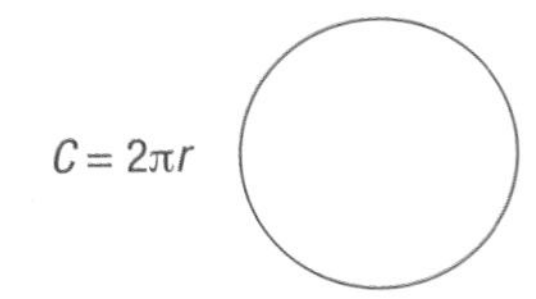

Worked example

Find the area and circumference of a circle of radius 4 cm.

Solution

Area $A = \pi r^2$
$= \pi \times 4^2$
$= 50.3\text{ cm}^2$ (three significant figures).

Circumference $C = 2\pi r$
$= 2 \times \pi \times 4$
$= 25.1\text{ cm}$ (three significant figures).

Tips for success

- Some questions tell you the diameter – so you must divide this by 2 to get the radius before starting to use the formulae.

Check your understanding 20.1

For each of these circles, find the area and the circumference, to three significant figures:

1. Circle with radius 3 cm.
2. Circle with radius 5 cm.
3. Circle with diameter 12 cm.
4. Circle with radius 8 cm.
5. Circle with diameter 14 cm.
6. Circle with diameter 7 cm.

Surface area and volume of a cylinder

Suppose a cylinder has radius r and height h.

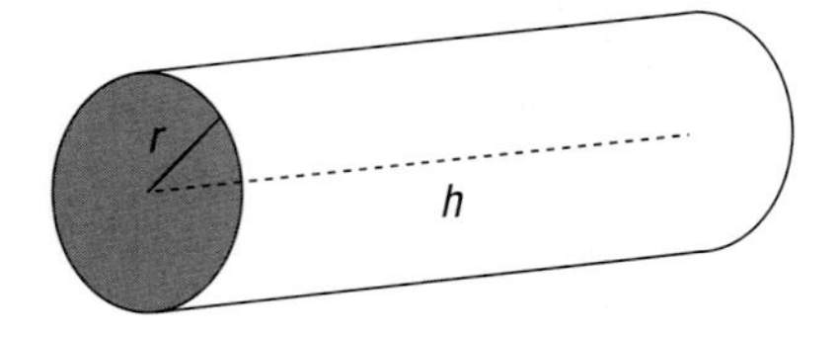

The **curved surface area** of the cylinder is given by $2\pi rh$.
The **volume** of the cylinder is given by $V = \pi r^2 h$.

Worked example

Find the curved surface area and the volume of the cylinder shown.

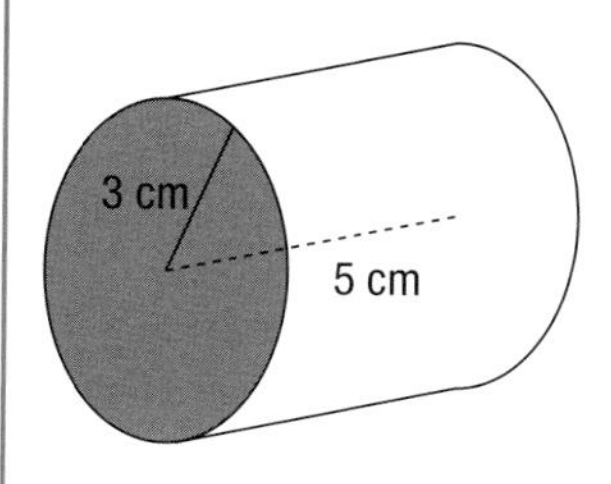

Solution

For this cylinder, $r = 3\,\text{cm}$ and $h = 5\,\text{cm}$.

Curved surface area $= 2\pi rh$
$= 2 \times \pi \times 3 \times 5$
$= 30\pi$
$= 94.2\,\text{cm}^2$ (three significant figures).

Volume $= \pi r^2 h$
$= \pi \times 3^2 \times 5$
$= 45\pi$
$= 141\,\text{cm}^3$ (three significant figures).

Check your understanding 20.2

For each of these cylinders, find the curved surface area and the volume, correct to three significant figures.

1 Cylinder with $r = 5\,\text{cm}$ and $h = 8\,\text{cm}$.

2 Cylinder with $r = 7\,\text{cm}$ and $h = 12\,\text{cm}$.

3 Cylinder with $r = 3\,\text{cm}$ and $h = 5\,\text{cm}$.

4 Cylinder with $r = 4\,\text{cm}$ and $h = 2\,\text{cm}$.

5 Find the volume of this cylinder.

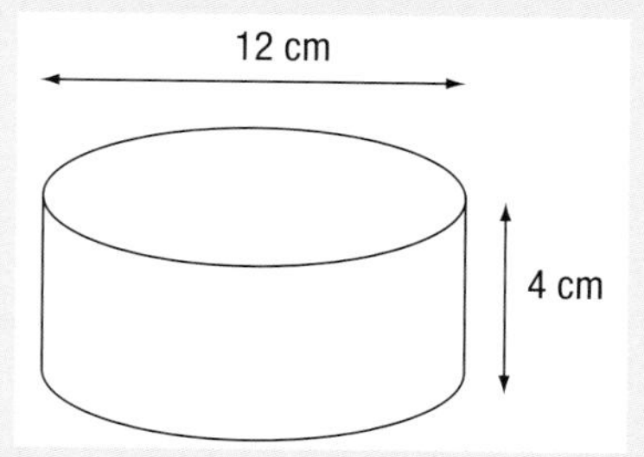

6 Find the total surface area of this cylinder. Include the two circular ends.

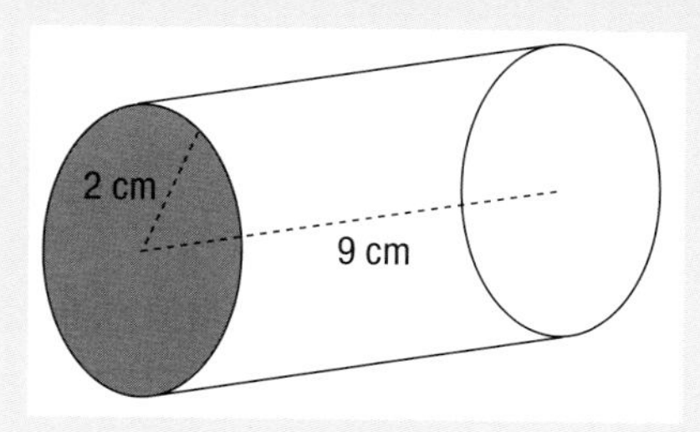

Prisms

In Chapter 16 on page 67 you found the surface area of a prism by drawing its net. To find the **volume** of a prism, simply multiply the cross-section area by the length.

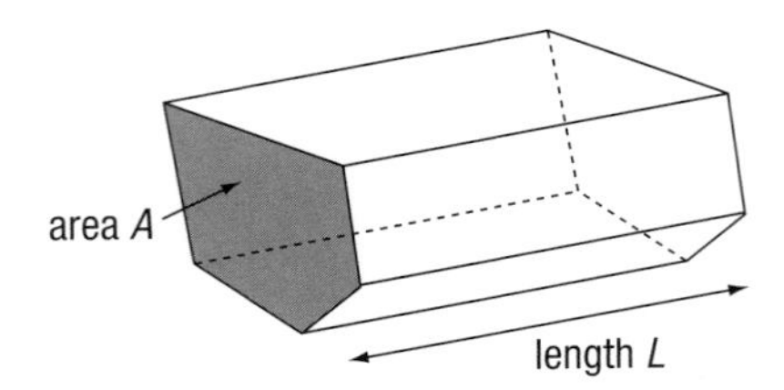

volume = cross-section area × length = $A \times L$

Worked example

Find the volume of this prism.

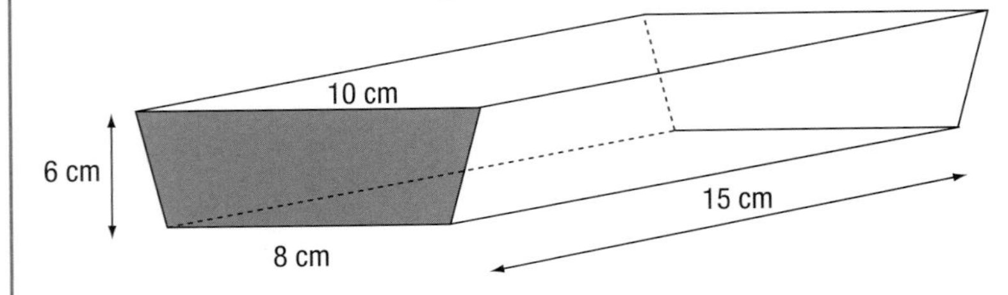

Solution

The shaded end is a trapezium, of area $A = \frac{1}{2} \times (10 + 8) \times 6 = 54\,\text{cm}^2$. So the volume is $54 \times 15 = 810\,\text{cm}^3$.

Tips for success

- As with all questions on area and volume, be sure to include correct units as part of your answer.

Check your understanding 20.3

Find the area of cross-section (shaded) for each prism, and hence work out the corresponding volume.

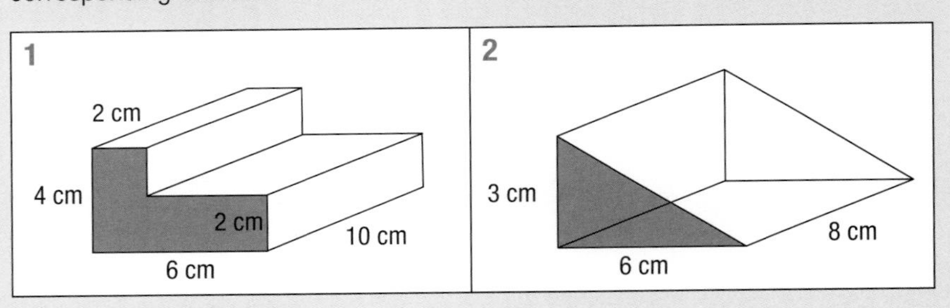

Spotlight on the test

1 The diagram shows a circle.

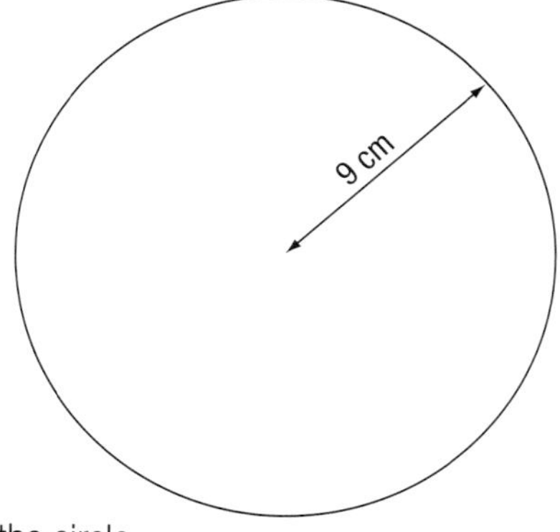

a) Calculate the circumference of the circle.
b) Calculate the area of the circle.
Give your answers correct to three significant figures. [2]

2 The diagram shows a semicircle.

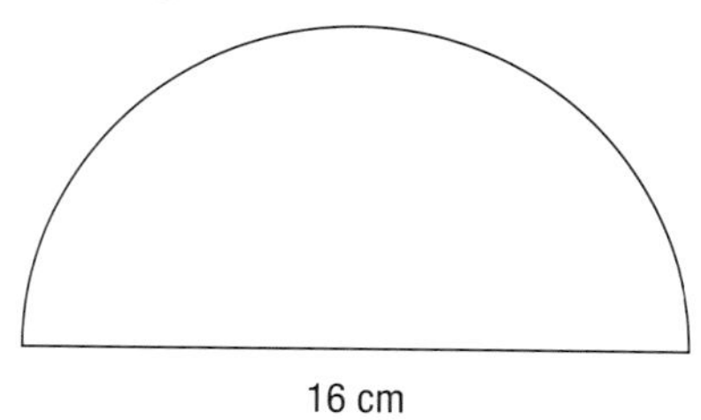

a) Write down the radius of the semicircle.
b) Find the perimeter of the semicircle.
Give both answers correct to three significant figures. [2]

3 The diagram shows a can of food.

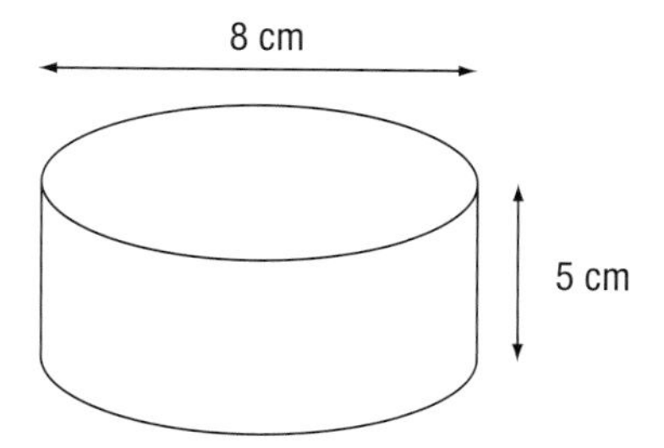

a) Find the volume of the can.
b) A rectangular label is placed around all of the curved part of the can. Calculate the area of the label. [4]

4 The diagram shows a triangular prism, whose cross-section is a right-angled triangle of sides 6 cm, 8 cm, 10 cm.

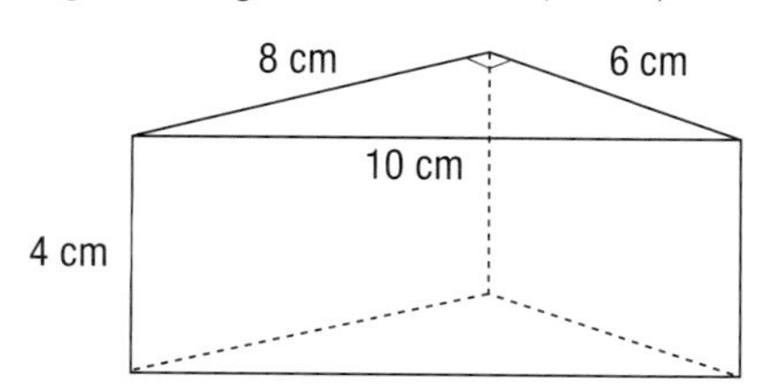

a) Calculate the area of the triangular cross-section.
b) Calculate the volume of the prism. [3]

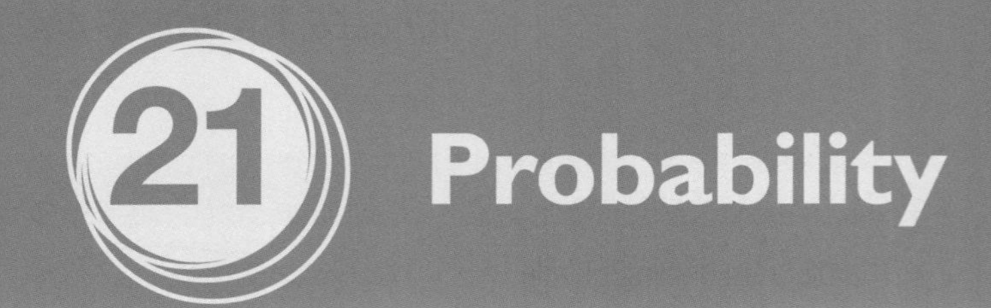

Probability

Basic probability

Student's book references

- Book 1 Chapter 23
- Book 2 Chapter 23
- Book 3 Chapter 23

Tips for success

- This definition of probability requires that each of the outcomes is equally likely.

Tips for success

- You could have written this answer as a decimal – 0.15 – but it is perfectly acceptable just to leave the answer as a fraction.

The **probability** of an event occurring is measured by a number between 0 and 1. An **impossible** outcome has a probability of 0; an outcome which is **certain** has probability 1. **Theoretical probability** can be calculated using the equation:

$$\text{Probability} = \frac{\text{number of ways of getting required outcome}}{\text{total number of all possible outcomes}}$$

Worked example

A bag contains ten red balls, seven green balls and three blue balls. A ball is chosen at random. Find the probability that it is blue.

Solution

The total number of balls is $10 + 7 + 3 = 20$. Each ball is equally likely to be chosen. The number of blue balls is 3.

So probability (choosing a blue) $= \frac{3}{20}$.

A probability experiment may have a number of different, non-overlapping results – a ball may be red, green or blue, for example. These are called **mutually exclusive** outcomes, and their probabilities will always *add up to 1.*

Worked example

When Boris plays chess with a certain friend the result is either that he wins, draws or loses. The probability that Boris wins a game is 0.45 and the probability that he draws is 0.3. Find the probability that Boris loses a game.

Solution

Let the probability that Boris loses be p.

Then $0.45 + 0.3 + p = 1$

$0.75 + p = 1$

$p = 1 - 0.75$

$p = 0.25.$

Probabilities may be used to estimate relative frequencies, and vice versa, as in this next example.

Worked example

When I toss a biased coin the likelihood of getting 'heads' is 0.35. The coin is tossed 20 times. Estimate the number of times that 'heads' occurs.

Solution
Estimated number = $20 \times 0.35 = 7$ times.

Check your understanding 21.1

1 A spinner has four equal sectors, labelled 1, 2, 3, 4. The spinner is spun once. Find the probability that the score shown is 3.
2 A spinner has five equal sectors, labelled 1, 2, 2, 3, 3. The spinner is spun once. Find the probability that the score shown is 3.
3 In a bag there are 100 toy building bricks. Twelve of them are yellow. Jim selects one brick. Work out the probability that his brick is a yellow brick.
4 When I reach a set of traffic lights on the way home they are either red, amber or green. The probability that they are red is 0.4 and the probability that they are green is 0.5. Find the probability that they are amber.
5 In a simple archery game each shot scores either 5, 2, 1 or 0 points. The probabilities of these outcomes are shown in the table below. One probability is missing from the table.

Score	5	2	1	0
Probability	0.1	0.2		0.3

a) Work out the probability of getting a score of 1.
b) Work out the probability of *not* scoring 2.
c) Twenty shots are taken. Estimate the number of times that 5 is scored.

Sample space diagrams

Suppose an experiment is being conducted twice. We can show the combined outcomes in a sample space diagram.

Worked example

Two fair dice are thrown and the total is recorded. Find the probability that the total is 10.

Solution
Draw up a table, and then write in the totals.

	1	2	3	4	5	6
1						
2						
3						
4						
5						
6						

	1	2	3	4	5	6
1	2	3	4	5	6	7
2	3	4	5	6	7	8
3	4	5	6	7	8	9
4	5	6	7	8	9	10
5	6	7	8	9	10	11
6	7	8	9	10	11	12

We see that a total score of 10 occurs in 3 of the 36 places within the body of the table. So, probability (total = 10) = $\frac{3}{36} = \frac{1}{12}$.

Check your understanding 21.2

1 Two coins are thrown, and for each one the outcome (heads, H, or tails, T) is recorded.

a) Copy and complete this sample space diagram to show all four outcomes.

	H	T
H	(H, H)	
T		

b) Work out the probability of getting double heads.

2 Two dice are thrown, and the **product** of their scores is recorded.

a) Draw a sample space diagram to show all 36 outcomes.

b) Work out the probability that the product is 12.

Spotlight on the test

1 A bag contains 100 tiles with letters printed on them. Forty of the tiles have a vowel, the rest consonants. One tile is chosen at random.

a) Work out the probability that it is a vowel.

b) Work out the probability that it is a consonant. [2]

2 A stamp is chosen at random from my stamp collection. The probability that it is French is 0.2. Work out the probability that it is *not* French. [1]

3 A bag contains 20 coloured balls. The probability of choosing a red ball is $\frac{2}{5}$.

a) Work out how many red balls are in the bag.

Some more red balls are now added to the bag. The probability of choosing a red now increases to $\frac{1}{2}$.

b) Work out how many red balls have been added. [2]

4 A box of breakfast cereal contains a free toy, which is one of a plane, a ship, a train or a car. The table shows the probability of getting each toy. One probability is missing from the table.

Toy	Plane	Ship	Train	Car
Probability	0.15	0.35		0.3

a) Work out the probability that a box contains a train.

b) Work out the probability that a box does *not* contain a ship.

c) Marina buys 20 boxes of breakfast cereal. Estimate the number of toy cars that she will obtain. [3]

5 A fair spinner with the numbers 1, 2, 3 and a fair die with the numbers 1, 2, 3, 4, 5, 6 are thrown. The scores are added together. Work out the probability that the total is 8. [3]

Written and mental arithmetic methods

Long multiplication

Student's book references

- Book 1 Chapter 23
- Book 2 Chapter 23
- Book 3 Chapter 23

Sometimes you need to multiply two numbers by longhand. You may know a traditional columns method for doing this. Two other common methods are the grid method and the gelosia method.

Worked example

Work out 146 × 27 without using a calculator.

Solution

Grid method:

100	40	6	×
			20
			7

100	40	6	×
2000	**800**	**120**	20
700	**280**	**42**	7

2000 + 800 + 120 = 2920

700 + 280 + 42 = 1022

So 2920 + 1022 = 3942

Gelosia method:

6 × 7 = 42, write this off-set as shown

1	**4**	**6**	×
			2
		4/2	**7**

Then complete the other boxes:

1	**4**	**6**	×
0/2	0/8	1/2	**2**
0/7	2/8	4/2	**7**

Then, add down the diagonals, carrying where needed:

	1	**4**	**6**	×
	0/2	0/8	1/2	**2**
	0/7	2/8	4/2	**7**
3	**9**	**4**	**2**	

So 146 + 27 = 3942

Tips for success

- Long multiplication with decimals: you can multiply as if the numbers were whole numbers. Then locate the decimal point by counting how many figures come after the decimal points in the original question.

Worked example

Work out 17.2 × 0.23 without using a calculator.

Solution

Using one of the standard written methods, we get 172 × 23 = 3956. The original problem, 17.2 × 0.23, has three figures after decimal points. So will the answer, hence 17.2 × 0.23 = 3.956.

Check your understanding 22.1

Use any written method to find these answers

1 132 × 46	**2** 18 × 59	**3** 243 × 19	**4** 88 × 64
5 337 × 56	**6** 203 × 11	**7** 442 × 66	**8** 365 × 24
9 3.14 × 16	**10** 8.2 × 7.4	**11** 9.9 × 18	**12** 17.4 × 2.5
13 3.6 × 0.4	**14** 0.4 × 0.3	**15** 0.24 × 680	**16** 4.3 × 3.4

Rearranging calculations

Knowing the answer to a calculation allows you to deduce the answer to a similar one without working it all out in full, as in this example.

Worked example

You are told that

$$\frac{42.8 \times 3.9}{7.5} = 22.256$$

Use this result to work out the values of

a) $\dfrac{4.28 \times 390}{0.75}$

b) $\dfrac{42\,800 \times 39}{22\,256}$

Solution

a) $\dfrac{4.28 \times 390}{0.75} = \dfrac{4.28 \times 3.90 \times 100}{7.5 \div 10} = 22.256 \times 100 \times 10 = 22\,256$

b) $\dfrac{42.8 \times 3.9}{22.256} = 7.5$ and so $\dfrac{42\,800 \times 39}{22\,256} = 7.5 \times 1000 \times 10 \div 1000$
$= 7.5 \times 10 = 75$

Spotlight on the test

1 You are given that

$$\frac{126 \times 55}{154} = 45$$

Use this result to work out the answer to the following calculations:

a) $\dfrac{12.6 \times 5.5}{15.4}$ **b)** $\dfrac{15.4 \times 45}{1.26}$ [2]

2 You are given that

$7 \times 25\,784 \times 10 = 1\,804\,880$

Use this result to work out the answer to

$1\,804\,880 \div 25\,784$ [1]

3 Work out 42.3 × 2.4 without using a calculator. Show all your working. [2]

23 Problem solving

P1 A ball of string weighs 72 grams. Kieran cuts a piece 50 cm long and weighs it on an electronic scale, and finds it weighs 3 grams. How long was the original ball of string? Give your answer in metres. [3]

P2 Here are some clues about five whole numbers:

- the median is 6
- the mode is 5
- the range is 3.

Find the five numbers. [2]

P3 Five snails are training for a race.
Snail A travels 4 cm in 5 seconds.
Snail B travels 6 cm in 8 seconds.
Snail C travels 7 cm in 10 seconds.
Snail D travels 17 cm in 20 seconds.
Snail E travels 24 cm in 40 seconds.
Work out which snail travels the fastest. [2]

P4 The diagram shows a circle and a rectangle. The circle has the same area as the rectangle. Calculate the width W of the rectangle. Give your answer to three significant figures.

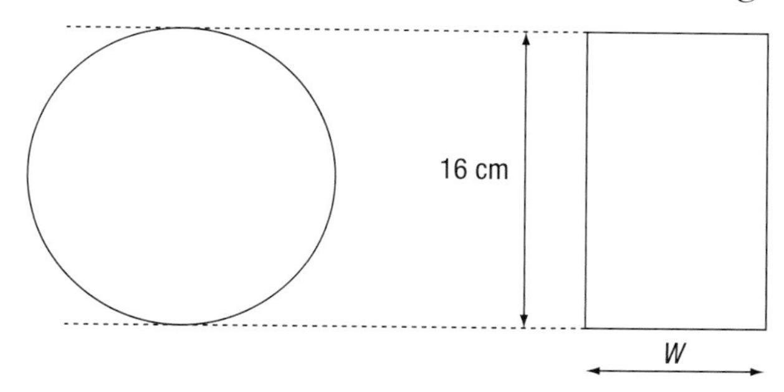

[2]

P5 The numbers a, b, c, d, e have a mean of 14.6 and a range of 6.

a) Find the mean of $a + 2$, $b + 2$, $c + 2$, $d + 2$, $e + 2$.
b) Find the range of $a + 2$, $b + 2$, $c + 2$, $d + 2$, $e + 2$. [2]

P6 A bag contains some coloured beads.
Each bead is red, green or blue. A bead is chosen at random. The probability that the bead is red is double the probability that the bead is green. The probability that the bead is green is three times the probability that the bead is blue. There are fewer than 20 beads in the bag. Work out the total number of beads in the bag. [2]

P7 The diagram, which has not been drawn accurately to scale, shows an equilateral triangle BCP.

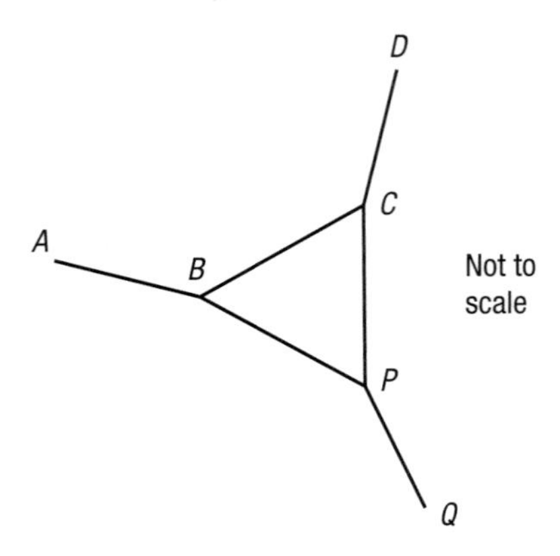

The points A, B, C and D lie at the vertices of a regular polygon with ten sides. The points D, C, P and Q lie at the vertices of a regular polygon with n sides.
Work out the value of n. [4]

P8 At Underbridge School the ratio of boys to girls is exactly 5 : 6. The head teacher cannot remember the exact number of pupils at the school, but she knows it is either 235 or 253. Work out how many boys there are at Underbridge School. [2]

P9 A circular running track has a circumference of 100 metres. Work out its diameter. Give your answer to three significant figures. [2]

P10 The three sides of a triangle are 18 cm, 30 cm and 35 cm. By calculation decide whether this is a right-angled triangle. Show all of your working. [3]

P11 Here is the first part of a number sequence:
17, 21, 25, 29, 33, …
Which of these numbers will appear somewhere in the sequence?
64, 69, 71, 101, 1037 [3]

P12 The angles in a triangle are $x°$, $2x + 10°$ and $3x - 70°$.
a) Write this information as an equation in x.
b) Find the three angles in the triangle.
c) What special name is given to this type of triangle? [4]